University of Dublin
Division of In - Service Education.

University of Dublin
Division of In - Service Education.

Curriculum development in mathematics

Curriculum development in mathematics

GEOFFREY HOWSON
Reader in Mathematical Curriculum Studies, University of Southampton

CHRISTINE KEITEL
Senior Research Fellow, Institute of Mathematical Education, Bielefeld

JEREMY KILPATRICK
Professor of Mathematics Education, University of Georgia

CAMBRIDGE UNIVERSITY PRESS
Cambridge
London   New York   New Rochelle
Melbourne   Sydney

Published by the Press Syndicate of the University of Cambridge
The Pitt Building, Trumpington Street, Cambridge CB2 1RP
32 East 57th Street, New York, NY 10022, USA
296 Beaconsfield Parade, Middle Park, Melbourne 3206, Australia

First published 1981
First paperback edition 1982

Printed in the United States of America

British Library Cataloguing in Publication Data

Howson, A.G.
Curriculum development in mathematics.
1. Mathematics – Study and teaching–
   History – 20th century
2. Curriculum planning – History
I. Title II. Keitel, Christine
III. Kilpatrick, Jeremy
510′.7′1   QA11   80–41205

ISBN 0 521 23767 X hard covers
ISBN 0 521 27053 7 paperback

CONTENTS

University of Dublin
Division of In - Service Education.

PREFACE

During the 1960s and 1970s there were few, if any, countries which did not attempt radically to reform school mathematics curricula. This reform movement can, indeed, be viewed as the greatest experiment ever to take place within mathematics education. Of course, it was not conducted upon classical research lines: there were, in general, no control groups, hypotheses were often implicit, and evaluation, when it did take place, was frequently carried out in a haphazard manner. In the mid 1970s the curriculum development boom was to end almost as rapidly as it had begun.

What are the lessons to be learned from these two decades of frenzied activity? How can success be attained and mistakes avoided in the future? What, indeed, is the future for curriculum development?

These are key questions which mathematics educators everywhere must answer. We hope that this book will help them in their task. What we have attempted to do is to set curriculum development in mathematics in both a historical and a more general social and educational context. We have sought not only to give a descriptive account of developmental work, but also to go beyond mere description in an attempt to provide a theoretical base for criticism and analysis.

We hope that the book will prove of interest and value to all those engaged in mathematics education, for everyone should be aware of his responsibilities so far as development is concerned.

The exercises form a special feature of the book. They are intended primarily to assist those who wish to use the book in conjunction with post-graduate, in-service or other courses. However, they do form an integral part of the text and we hope that all readers will pay due attention to them. The problems posed cannot be readily solved. Most, however, can be interpreted at a variety of levels; many may well form the starting point for discussion, further reading, dissertations, theses or general research work. All should help the reader to appreciate more clearly the complexity of the task in which we are engaged.

We have had much help and encouragement during the writing of this text from a variety of sources: in particular, our spouses, colleagues and students. We are indebted and grateful to them all.

January 1980

Geoffrey Howson
Christine Keitel
Jeremy Kilpatrick

1

Curriculum Development: an Introduction

This century has seen vast changes in school systems everywhere and in the education they offer. For example, in the developed countries, secondary education for all has become a reality; elsewhere, rapid progress to that end is being made. However, not only is education being offered more widely, but it now has different goals. Changes in the social and economic structures of society have had profound implications for education, as have the growth of new technologies and of knowledge. Such changes will continue to occur and to present challenges to the educator and, in particular, to the curriculum developer. The need for curriculum development will not be transient.

In this book we describe some of the features of curriculum development as it affects mathematical education. We shall look at the way that curriculum development has taken place in the past – especially within the last two decades; the various forces that have influenced the form that it has taken and the successes it has achieved; the management procedures which have been devised; and the attempts made to evaluate its outcomes. Finally, we look critically at the reform period in retrospect, its achievements and failures, and the lessons to be learned. We do this in an attempt to ensure that in future curriculum development will proceed with greater ease; that proposed innovations will be examined more critically and that, when it is decided to implement changes, those changes will be effected more successfully – that they will not only reach classrooms, but will do so in an ungarbled form reflecting the originators' aims and objectives. We hope also that this book will play a part in increasing the professional competence of the teacher by allowing him to see his role in curriculum development set in a wider context: that it will make a contribution to the education of 'informed decision-makers'.

It is our aim in this first chapter to identify certain key features of curriculum development and to introduce several ideas to be developed later. Occasionally, we shall attempt to provide 'classifications'. Notwithstanding their crudeness, and the way in which they, often misleadingly, bring apparent clarity and simplicity to what are, in fact, extremely complex situations, we hope they will provide the reader with a framework that will help him better to consider and understand what follows.

First, however, we must consider more fully what is meant by the word 'curriculum'. Many attempts have been made by educationists to define the term in an intensive way (see, for instance, the examples quoted in Hooper (1971) and Becher & Maclure (1978)). Unanimity has not been forthcoming. Yet it is now generally accepted that the word can no longer merely signify 'the syllabus', whether that is drawn up by a central organisation or within a single school. Certainly the restricted outlook which views curriculum development as merely the production of new syllabuses and texts must almost certainly lead to failure and disappointment. If curriculum development is to succeed, then one must hold a wider view of the curriculum than this and plan changes to take into account all of the factors involved. Neither content nor method can be viewed in isolation, and both can be planned only when clear aims are envisaged both for education in general and mathematical education in particular. It will be necessary also to consider the means by which the course and the students taking it can be assessed. The best plans will come to nought unless examination systems test desired objectives and encourage (rather than act against) the attainment of educational and mathematical goals. Curriculum, therefore, must mean more than syllabus – it must encompass aims, content, methods and assessment procedures. One cannot truly talk, then, of a 'national curriculum', for it depends upon individual teachers, their methods and understanding, and their interpretation of aims, guidelines, texts, etc. The part played by the individual teacher must, therefore, be recognised.

1. Pressures that can serve to initiate curriculum development

We have already mentioned some of the forces which prompt innovation. Such pressures can come from various quarters and it is of value to identify some of these. Here, as elsewhere in this book, a list is supplied with the warning that any such classification is arbitrary in the sense that other educators might well opt for alternative presentations. Moreover, rarely will it be found that a particular innovation will fit neatly into any one category. Normally there will be a mixture of pressures, a variety of barriers to be overcome, etc. The problem of curriculum development is to ensure that a response is made to *all* the relevant pressures, and attempts made to surmount *all* the barriers. Many innovations have failed because this basic fact has not been recognised.

1.1.   *Societal and political*

Of all the pressures that initiate curriculum development none is greater than that exerted by society.

Economical and technological development has not only meant that society *can* provide more education, but that it *must* do so. Moves towards egalitarian societies have necessitated, and been affected by, changes in educational practice. Education has increasingly been seen as a means through which existing systems of values can be changed (or, in some cases buttressed). Employment demands have altered with important consequences for education.

It is as a result of societal pressures that secondary education is no longer a training solely for the élite and that in many countries it is organised on comprehensive lines; changes to which mathematics educators have been forced to respond.

Such pressures may be, but are not necessarily, 'national'; often they may be local or restricted to a particular section of the population. Thus particular social pressures in the London educational authority, which caused amongst other things an extremely high rate of teacher-changeover, led to a project, SMILE,* designed especially to meet local demands. Again, in the USA, Project SEED has concentrated on the problems of the educationally disadvantaged.

Occasionally the demands of society can be quite specific: to prepare children for a change in currency (as happened in Britain in the early 1970s) or for metrication. The latter innovation has led to difficulties in Britain because, for once, the schools have responded to society's demands faster than society itself could change. Thus we now hear complaints from industrialists that their young employees know only the metric system whereas their machines and the problems they cause are

*A glossary of acronyms is to be found on p. 268.

still 'imperial'. The need to keep changes in society and education 'in phase' is clearly demonstrated. Similar considerations would, for instance, militate against the introduction of informal teaching based on a questioning approach in a society still organised like many industries on rigid, patriarchal lines: independence of thought is not an aim of education in such societies.

1.2. *Mathematical*

Mathematics, too, has grown phenomenally this century. Nineteenth-century mathematics is now viewed in a new perspective emphasising structural considerations, and new branches of the subject have mushroomed. The computer has caused changes of emphasis; for example, numerical mathematics has been revolutionised, and mathematics has found a wide variety of new applications as disciplines such as economics and geography have become more and more quantitative. More recently, work by the French mathematician René Thom and others – catastrophe theory – has opened up exciting new possibilities in biology and other fields.

Such developments have resulted in a complete revision of the content of university courses and their influences have also been felt in schools. In addition, changes at one level of education will impose another pressure on lower levels, the need to prepare students for a new approach to the subject. More generally, changes at one level of education are almost certain to cause problems at the relevant interfaces, whether these are between two different sectors of education, or between the educational system and employers.

1.3. *Educational*

Pressures for change can arise from within the educational system as a result of research, new educational theories, or the pioneering developmental work of individuals. Thus the work of Piaget has caused educators to reconsider some of the curricular aims of the early school years, while that of Bloom and his co-workers has greatly affected the way in which many educational problems have been approached (see chapters 5, 6 and 7).

New technologies have been applied to education. In practical terms this has meant that new aids and apparatus have been made available: the hand calculator, the microprocessor, the video-tape-recorder. In theoretical discussion attempts have been made to transplant procedures, language and models from other disciplines into education, in particular, the 'systems approach' of engineering.

1.4. *The rewards of innovation*

The desire to change, to act as an innovator, to explore new areas is strong in most intelligent people. Many educators, then, welcome the opportunity to deviate from practices that are becoming routine. Innovation is exciting, attracts others' attention to one's work and often brings professional advancement. Moreover, the educator is attempting to solve an insoluble problem: for if a goal is attained it will almost invariably be replaced by a more ambitious one. There is a constant spur to change.

Yet another incentive to innovate is provided by the example of others. Countries and individuals fear that they will be left behind, that their ways are becoming obsolete. There are also commercial pressures for change as publishers and authors see new markets.

Exercises

1 Investigate how in this century the numbers of students enrolled for primary, secondary and higher education have grown in your country. What differences likely to affect mathematical education have occurred in patterns of employment?

2 It can be argued that someone 'loses' as a result of every innovation, be they particular types of pupils, teachers, or taxpayers. Investigate this assertion and consider its truth in the context of particular projects.

3 Discuss some of the problems that can arise at the 'interfaces' as a result of curriculum development.

4 'One fancies, indeed, that experiments in education would not be necessary; and that we might judge by the understanding whether any plan would turn out well or ill. But this is a great mistake. Experience shows that often . . . we get quite opposite results from what we had anticipated.' (Kant, *On Pedagogy,* 1803.) Is Kant's thesis valid today? If you believe the answer is 'Yes', then provide examples to substantiate your claim.

5 'Many educational changes are at least accelerated if not caused by wars.' (Mallinson, 1966.) Investigate and comment.

6 Modern, 'Bourbaki'-inspired mathematics has now been absorbed into university mathematics. What new trends can be discerned in undergraduate courses?

7 Investigate and describe the impact on curriculum development of the increased access of those in schools to sophisticated reprographic

facilities, such as small litho machines, heat duplicators and electronic stencil makers.

2.   **Barriers to curriculum development**

As we have seen there are many pressures which serve to initiate curriculum development. Yet radical changes in the curriculum occur somewhat infrequently and many attempted changes fail to disturb the equilibrium that would appear to exist in the classroom. One is almost tempted to believe that there is a curricular analogue to Newton's Third Law 'To every action there is an equal and opposite reaction'. This would be to take too pessimistic a view. Yet there do exist clear barriers to change which must be overcome. A knowledge of these barriers can only assist in the planning of successful development. In the succeeding chapters we shall see how attempts have been made to surmount these obstacles. For the moment we seek only to indicate their nature. The list of 'types' which we give is due to Dalin (1978). We observe that, in practice, the first two types are difficult to separate: it can be argued that all 'values' reflect social and political interests and *ipso facto* are concerned with 'power'.

2.1.   *Value barriers*

Different people have different ideologies, interests, ... These differences can arise from a variety of causes, political, religious, educational, social background, etc, and will significantly affect the way in which people react to suggested innovation. The effects are perhaps most readily observed on broad educational issues such as comprehensivisation, the raising of the school-leaving age, selection or open access to higher education. Yet their effects can also be significant within mathematical education. Reactions to mixed-ability teaching, for example, are largely governed by one's sense of values. So is the extent to which school mathematics is seen as a service subject, there to satisfy the needs of science teachers and would-be employers. Again, many teachers find it abhorrent to teach 'cook-book' statistics to biologists, economists, etc.; to encourage the use of mathematics that has not been 'understood' or 'proved' affronts their sense of values.

2.2.   *Power barriers*

Significant innovations are often accompanied by shifts in the balance of power. In Britain there is currently talk of the imposition of a 'core curriculum', something which would remove some of the teachers' powers and reinforce those of central government. In some countries there has been opposition to newly-established curriculum development centres and institutions because these have been seen as shifting the balance of power, sometimes away from the teacher, sometimes away from central authority. Universities have fought and are fighting to retain their influence on the school curriculum; parents, students and employers wish to exercise more power.

2.3. *Practical barriers*

Whether or not a new branch of pure mathematics, say, category theory, should be taught in schools is at first sight a question of values. If, however, a central administration decided to include category theory in the school curriculum it would soon be faced by a different type of barrier. The practical barriers are: few, if any, teachers know anything about category theory; it would be extremely expensive and perhaps logistically impossible to mount an in-service teacher education programme to remedy this deficiency; additional money and manpower would be needed to produce the necessary textbooks and new revised examinations; to make room for category theory, some traditional topic would have to be omitted; and that it would be extremely difficult to demonstrate why (and how) such a change was in the general interest.

Similar considerations will apply to other, less contentious innovations. Thus the recommendation by the Dainton Committee (1968) that 'normally (in England) all pupils should study mathematics until they leave school, and only in exceptional circumstances should it be held to be possible or desirable for a pupil to opt out' stood no chance of being implemented on practical grounds. There was a chronic shortage of qualified mathematics teachers and, hence, no possibility of providing the new courses (in any worthwhile form).

2.4. *Psychological barriers*

People are often happiest doing what they are used to doing (*cf.* Fey, 1979). The known provides a degree of security; innovation means risk-taking. There will often be psychological barriers to be overcome before an innovation is accepted.

These barriers can take a variety of forms; one of the most difficult to surmount being a previous unsuccessful innovatory experience. 'Once bitten, twice shy' is very much true of curriculum development. An ill-conceived, ill-executed change can be the cause of many adverse effects, not the least of which is to create opposition to further innovations.

Another barrier concerns incentives. If individuals or groups can see how they stand to gain by innovation, then changes in behaviour are more easily accomplished. Here 'gain' can take many forms: change can support one's values, increase one's power or assist in the fulfilment of personal goals. In the case of the innovator it is usually possible to identify the incentives. But where are the incentives for the ordinary teacher asked to adopt the innovation? Lack of incentives can be a powerful barrier to curriculum development.

Exercises

1 Think of more examples of how values affect attitudes to curriculum change.

2 Who exercises control of the mathematics curriculum in the schools of your country? How much control has (a) the head of the school, (b) the head or chairman of the mathematics department (if such a position exists)? What influence do the universities have, and how do they exercise it? Has the balance of power changed in the last decade? What forces, if any, are attempting to change that balance?

3 'Teachers should be allowed to carry on doing what they do well.' Discuss.

4 As an example of a 'value' barrier we have quoted the reluctance of many mathematicians to teach 'cook-book' statistics. Comment on the argument that this is, in fact, a 'power' barrier, that 'teaching for understanding' retains 'power' for mathematical initiates and means that the mathematician's job cannot be usurped by, say, the biology teacher. Describe and discuss similar examples, such as the problem of who should teach mathematics to engineering undergraduates: engineers or mathematicians?

3. Types of curriculum development

So far we have described the need for, and the barriers opposing, curriculum development but have not written of the ways in which developments can take place. Innovation in mathematical education is often thought of in terms of the work of major projects such as the School Mathematics Study Group (SMSG) in the USA or the School Mathematics Project (SMP) in England. However, if misconceptions are to be avoided, it is essential to realise that curriculum development does not solely comprise the efforts of such bodies.

It is, of course, natural that one should hear most of those initiatives that are meant to apply throughout a school system. In countries with a centralised system (see chapter 4), this may be the only recognised form of curriculum development. The outcomes in the form of syllabuses and/or materials will then be imposed on all schools. In other countries changes will be proposed (and materials produced) in the hope that they will be widely used (e.g. by the SMSG, the SMP, or the SSMCIS (see chapter 3)). In both cases the overwhelming majority of teachers asked to adopt the innovation will have played no part in the planning, goal-fixing, writing or testing. The problems of dissemination and implementation (i.e. of making teachers aware of the aims and objectives of the innovation, and of helping them to accommodate to changes in content and method) will, therefore, loom large.

Yet initiatives can be more local in nature, as our description of the Fife Mathematics Project (chapter 3) shows. A group of schools in a region, or those comprising a small educational system, combine, often under the guidance of a full-time 'professional' curriculum developer, to innovate. A distinguishing feature is that all the schools, and representative teachers from them, will be involved in decision-taking and in the developmental stages. Problems of dissemination and implementation are eased although, these will, of course, increase if attempts are made to encourage the use of published materials in schools other than those originally involved.

Finally, a single school or individual teacher can be our 'unit' for innovation. Two teachers may cooperate on team-teaching, or an individual may prepare a set of workcards to cover a particular topic. Problems of dissemination and implementation vanish.

Innovations can, of course, occur at all levels of education from kindergarten to university post-graduate level and with all types of learner in mind. It is obvious though that the pressures exerted, the barriers to be overcome and the modes utilised will differ considerably according to the age and ability levels of the students and the qualifications (often related) of the teaching staff.

Thus at a university level where the teachers' mathematical qualifications are high and individual institutions have a high degree of autonomy, development is usually in the 'individual' mode. Larger scale projects rarely arise and when they do (e.g. the (USA) Committee on the Undergraduate Program in Mathematics) the outcome is usually a set of guidelines to be interpreted by individuals rather than an exhaustive collection of classroom materials. The greater mathematical expertise of the teacher allows him to respond more quickly to mathematical pressures. (Yet he may, for a variety of reasons, be slower to respond to those exerted by society and become disastrously out of touch with the real learning needs of his students.)

At the other extreme, the primary school teacher is often lacking in mathematical knowledge; indeed she might well fear or hate the subject. Guidelines will hardly suffice to bring about worthwhile changes, and more help will be needed. Yet the vast numbers of teachers involved will present problems of dissemination and implementation which militate against a 'large-scale' approach, whilst 'local' action may be handicapped by the inability of many teachers to take decisions concerning aims, content and method from a position of mathematical strength. Such facts must be recognised if developmental work is to be successful, but they cannot be allowed to become the rationale for stagnation.

Exercises

1 Describe some 'small-scale' curriculum development work in which you have been involved as student or teacher.
2 Give reasons why university teachers may be slow to respond to pressures on the curriculum emanating from outside mathematics. Is this something to be deplored, or is it the case that it ensures 'academic stability' and provides a needed conservative presence in education?
3 Take various curriculum initiatives with which you are familiar and see if they fit naturally into the classification given. Are there changes you would wish to make to the classificatory system?
4 'Whatever is attempted without previous certainty of success, may be considered as a project, and amongst narrow minds may, therefore, expose its author to censure and contempt; and if the liberty of laughing be once indulged, everyman will laugh at what he does not understand, every project will be considered as madness and every great or new design will be censured as a project.' (Samuel Johnson, quoted in Ziman, 1976, p. 180.) Discuss.

4.   Strategies to be employed

The various strategies that are employed in effecting innovation have been analysed by several authors. (Summaries can be found in, for example, Hooper (1971) and Dalin (1978).) As we shall see when we come to consider projects in greater detail, innovators are usually constrained in their choice of strategies and may make use of a mixture of possibilities rather than any single one.

One obvious strategy is that employed by governments and other authorities when they issue fiats to the effect that 'from 1st October 19. . all schools in the system shall . . .'. Here an educational authority (local or national), a head teacher or a head of department uses its/his authority to impose an innovation on subordinates: a 'power-coercive' strategy. It is the way in which modern mathematics was brought to France and Federal Germany, and the way in which it was removed from Malawi and Nigeria!

Another coercive strategy employs not authority, but 'pressure'. Arguments such as 'if you cared for your pupils, . . . if you were up-to-date, you would be . . .' are used. Less emotive and more tangible methods can also be used to 'encourage' innovation. Two methods that have been employed are having a higher pass rate for the 'modern' syllabus in the centrally-controlled public examinations and paying higher salaries to those who teach the new courses.

Rather than relying on emotive pressures or tangible rewards attempts may be made to demonstrate to teachers the objective benefits that would follow from adopting the innovation. This strategy, the 'rational–empirical', is often linked with evaluatory exercises, for example, the claim (arising from the evaluation of the Swedish IMU Project (see chapter 3)) that 'pupils learn to take responsibility to a greater extent with IMU'.

Another, more fundamental, strategy for innovation is to attempt to change the attitudes, skills and values of those within the educational system in the belief that this will activate them to become innovators. This 're-educative' strategy is one which has been actively pursued by, for example, the Association of Teachers of Mathematics in Britain.

Generally speaking one finds that innovations are disseminated throughout centralised national or local educational systems by means of the two coercive strategies, whilst 'large scale' projects working in non-centralised systems make greater use of 'rational–empirical' arguments. 'Local' projects will tend to use the re-educative strategy in which in-service education and curriculum development are closely coordinated.

The type of strategy adopted is, of course, closely related to the teacher's standing within any educational system. In many cases the educational system effectively defines the strategy to be adopted. In general, the teacher's role and status within the system can be deduced from the strategy: whether he is to be treated as a servant of the system, a semi-autonomous being to be manipulated or urged along, or a professional capable of making sensible decisions. The teacher's initial acceptance of innovation and his commitment to it when difficulties arise will be crucially dependent on the type of disseminatory strategy employed.

Yet the teacher's role is not uniquely and clearly defined. It may well be, and indeed has proved to be, the case that the teacher's view of his role can differ from that which is assumed and assigned by the innovator. If the teacher visualises himself as the servant of the system, to be told what to teach (and, preferably, when), then he is unlikely to use successfully materials devised by innovators who see the teacher as an autonomous producer/director of what happens in the classroom. Thus, for example, it would appear that there is a mismatch between the view of the teacher's role held by many teachers using USMES materials and the view which the originators of the project would seem to hold (see chapter 7). Such mismatches can prove fatal to the successful implementation of innovation.

Exercises

1 Explain in more detail and with examples, why the type of disseminatory strategy adopted should 'serve to define the teacher's role within the educational system.'

2 Investigate teachers' reactions to difficulties created by curriculum development. Is it true that these are affected by the choice of disseminatory strategy?

3 As Head of Department you decide that the mathematical curriculum in your school should be changed. This view is not shared by two of your older staff – one in his forties, the other within five years of retirement. How would you handle this situation?

4 Describe the strategies that have been used to effect innovation in your country.

5. Who is involved in curriculum development?

Pressures to initiate curriculum development can come from many quarters, and various groups have an interest in educational innovation: politicians, educational administrators, teachers (and not only at the educational level immediately concerned), publishers, educational researchers, students, parents, employers, taxpayers etc. The way in which innovation proceeds will be very much influenced by the contribution of each of these groups to the decision-making process and by the way in which each group is called upon to respond to change.

How changes occur and the biases built into the innovations will greatly depend upon who is responsible for making the first moves. That the SMSG grew out of two conferences of university mathematicians, that the impetus for the SMP came from schoolteachers, and that reforms in France were largely initiated by outstanding professional mathematicians explains much of their later history, successes and failures. Some projects arose because educators not directly involved saw a need for them, others because teachers in the classroom wished to alter the *status quo*. This difference has often proved to be crucial in the dissemination process.

Yet an outstanding and worrying feature of many innovations is that initiation and subsequent decision-making has been unilateral: that only one group has been involved until too late, when, momentum having been attained, other interested parties have attempted to interfere.

Vital decisions concerning areas for reform have too often been made by subject-oriented power groups without regard to a country's overall educational aims. Yet who is to take decisions in a non-centralised system?

The construction of the school mathematics curriculum cannot be entrusted, say, to industrialists and other employers. Yet their voice must be heard and accorded weight. Those involved in curriculum design and development will have to take into account the views and needs of *all* interested parties, but these views will then have to be weighted and mediated. (The fact that certain parties are more articulate than others and can project their voices more loudly must always be remembered.)

The roles of various groups in curriculum development is a theme to which we shall return in chapter 4.

Exercises

1   Take particular examples of curriculum development and investigate (a) who initiated the innovation; (b) at what stage the views of other interested parties were gathered; (c) what changes, if any, resulted from such consultation.

2   Find examples of projects which were initiated by teachers and others initiated by 'educators not directly involved'. Contrast the ensuing patterns of dissemination.

3   Investigate the various ways in which attempts have been made to explain the aims and content of new mathematics programmes to parents. Discuss how you might attempt this explanation if you were

(a) a Head of Department in a school; (b) a project director; (c) a local inspector/adviser.

4 'Regardless of their many virtues, professional innovators know so much that they often have the effect of intimidating the people they are trying to help.' (Gruber & Vonèche 1977, p. 694.) Discuss.

6. The stages of curriculum development

We have already spoken of the initiation of innovative work and of its dissemination and implementation. We now draw together some of the ideas introduced in this chapter in an attempt to identify the various stages which can be discerned in curriculum development.

Innovation, as we have seen, begins with the *identification* of a need or a possibility, usually in response to a particular pressure (section 1). Next comes the *formulation* of a course of action. Once this has been done there will usually be a need to persuade other interest groups to *accept* the suggested innovation. Various actions may be necessary depending upon the mode of curriculum development (section 3) to be adopted. Perhaps a headteacher will have to be convinced, or it may be that schools and teachers must be found who are willing to participate, finance raised, examination boards persuaded to institute new examinations, universities and professional bodies to recognise the new qualifications etc. (section 5). There is a need for negotiation, a need which is likely constantly to recur.

Most innovations in education and, in particular, those we shall examine in this book, are concerned with the *development* of new materials and other physical resources. This is a key stage in most curriculum development projects and the one on which attention tends to be focused, usually to the neglect of the equally important, more costly and time-consuming, later stages. Such developmental work may well choose to build on existing research findings. In other cases research work in specific areas will have to be commissioned. During this stage, formative evaluation of the innovation will take place: the testing and assessing of the project's draft materials as they are used in the 'pilot' experimental institutions.

If an innovation is to be adopted throughout an educational system then it will be necessary to explain the aims to this wider audience. Many teachers will have to be introduced to the new materials and methods; in-service education will be needed. *Dissemination* has begun. Once the innovation has been accepted there is a need to help sustain it over a period of time, to provide an 'after-sales' service. We are now in the *implementation* phase. In this, as in earlier stages, there will be a need to consider and overcome potential barriers (section 2) to change, and to adopt a suitable strategy for propagation (section 4). In addition to the formative evaluation which occurs at the developmental stage, summative and comparative *evaluation* will also take place. This may be 'formal', for example, NLSMA (see chapter 7), or 'informal' in the sense of the accumulation of 'feedback' which led SMP to produce a booklet *Manipulative Skills in School Mathematics* which sought to emphasise some of the dangers and misunderstandings that had become evident at the implementation stage.

Such evaluation can, in fact, lead to the identification of new problems and so help initiate yet another round of innovation.

Exercises

1 Think of one or two changes you would like to see made within mathematical education. What problems of 'acceptance' would have to be overcome? What problems of dissemination and implementation would arise if the innovation was to be widely adopted?

2 Give examples of innovation that have reached some classrooms in a 'garbled' form. What further steps might have been taken to help prevent this?

3 An important question to ask of any innovation is 'Are the dominant forces for change internal (within the school, or the school system) or external?'. What effect is the answer likely to have on the later stages of development?

4 Take a selection of projects, locate and study articles written in the projects' early days by their originators, and hence attempt to identify the factors which led to the projects' establishment. Note any apparent later changes in policy and attempt to explain these.

2

The Historical Background

> More than once he went rejoicing abroad in foreign by-ways, led by
> love of wisdom, to see if he could find in those lands new books or
> studies which he could bring home with him.

These words of Alcuin (see Sylvester, 1970, p. 3) describe how
Ethelbert, his predecessor as schoolmaster at York Minister in the
eighth century, attempted to ensure that the curriculum at that school
was kept up-to-date. They emphasise how long a history curriculum
development has, and it is significant that two agencies for change are
identified, personal contact through travel and the book, which are still
highly influential today. Alcuin, himself, was to become a still more
renowned educator for he, the first member of England's 'brain-drain',
was to move from York to Aachen where he tutored the almost
illiterate Charlemagne and his court in a variety of subjects, including
mathematics, and led what came to be known as the Carolingian
renaissance.

Yet if we wish seriously to consider 'curriculum development' at York
then we shall need to have further information. Thus, for example, it
will be necessary to know by whom the school was established and for
what purpose; what views society held on education; what mathematics
was taught and used. Curriculum development, then as now, takes place
within a social, educational, political and administrative framework
owing much to history and geography. It is for such reasons that the
English School Mathematics Project could have no French analogue,
that reforms in the French style could not have taken place in the
Netherlands, and that the methods of the Dutch IOWO (Institute for
Development of Mathematics Education) cannot be transferred to the
USA.

1. **The development of educational systems and of the idea of progression**

Alcuin's school was, like all others existing in Western Europe at that time, an appendage of the Church. The education it provided was that needed by the Church's adherents, in particular, the clergy. The purpose of education was not, as in classical times, to produce a cultivated man of affairs, but an educated cleric: it was largely vocational. Learning was restricted to whatever fell within the Church's interests and doctrines, a similar situation to that which still exists today in many secular totalitarian states. In particular, interest in mathematics was rudimentary: in Ethelbert's case 'the different kinds of number and the various shapes' and sufficient astronomy to help determine the dates of such movable feasts as Easter. The books available were, for historical reasons, the derivative, uninspired works of Boethius rather than the Greek originals.

The individual schoolmaster had freedom to design his own curriculum, but freedom within a system that provided no opportunities for significant developments, one in which intellectual inquiry was directed backwards, towards classical learning, rather than forwards. The idea that the curriculum would have to change because of the progression of knowledge was still many centuries away.

The Church's domination of education was not to be challenged for many centuries, although with the growth of cities the schools did gradually become more secular in character. Schooling began, too, to fall into two types, the grammar schools in which Latin was the staple element in the curriculum and the humbler schools in which more emphasis was placed on the vernacular.

The universities, creations of the late twelfth and early thirteenth centuries, not unnaturally used Latin, a language to which they adhered for many years. The reasons for this were clear: Latin was the international language of scholars, that in which existing texts were written and indeed for many centuries that in which many key mathematical works, such as, Newton's *Principia,* Euler's *Introductio in Analysin infinitorum* and Gauss' *Arithmeticae Disquisitiones*, were to appear.

Even if this policy was inevitable, it nevertheless aggravated a position in which the mathematics of the people and that of the scholastic community drifted apart. The grammar schools which reflected the ideals of the academic world failed to meet obvious social needs. By the sixteenth century ingenious, but unlettered, artisans had invented the compass, gunpowder and printing and had sown the seeds of the idea of 'progressive' technology as an important part of life. Commercial and business life was increasing with an accompanying need for a new specialised arithmetic, invigorated by the growing adoption of the Hindu–arabic system of notation.

A new breed of superior craftsmen arose – Brunelleschi, Leonardo, Tartaglia, Stevin – and there was a growing demand for skilled navigators. An alternative, more utilitarian type of education was needed, one which the existing institutions were ill-equipped to provide. The need was mainly met by individuals: by mathematical practitioners, or Rechenmeisters who taught privately and published textbooks in the vernacular. A second tier of education began to arise.

The established system of education also experienced changes as a result of the Renaissance and the Reformation. Those that followed in the wake of humanists such as Erasmus tended to exacerbate the developing division. Greek and sometimes Hebrew, rather than mathematics, entered the curriculum of the grammar schools; the ancient civilisations were seen as models for the modern world and their writings as containing all the requisite knowledge for the 'educated man'. It was through the study of the latter that an aristocracy of intellect would be created. Alternative views on education were, however, expressed by Luther and other Protestant reformers. The Church's traditional hold on education had been destroyed or weakened and new aims for education could now be defined. Education could be seen as contributing to the prosperity of the city, to the efficiency of the home. Did it not demand the same financial outlay as roads or fortifications?

The effect of men such as Luther and Calvin was considerable. An early idea of education for the masses was planted in, for example, Protestant Germany, Massachusetts, Scandinavia and Scotland. Thus, in 1619 an attempt was made in Weimar to make education compulsory between the ages of six and twelve years, whilst a Scottish Act of 1696 decreed that it was the duty of every parish to provide a 'commodious house for a school' and a salary for a teacher.

Schools were, however, still governed by individuals, by cities, guilds and parishes. National education lay in the future. The curriculum depended on individuals and on local needs. Where local circumstances dictated that mathematics should be taught, for example, in the English seaports of Plymouth and Dartmouth, then a place was found for it in the curriculum. Elsewhere it was neglected. So, for example, in 1662, when he was a highly-paid civil servant in the Admiralty, Samuel Pepys, the English diarist and a graduate of Cambridge University had to engage a private tutor to teach him his multiplication tables. Shortly after, Pepys was elected one of the first Fellows of the Royal Society: a society whose full title 'The Royal Society of London for Improving Natural Knowledge' bore witness to the fact that the idea of 'scientific progress' had now become accepted. No longer could the curriculum remain static – at least in theory! – for knowledge was now recognised to be constantly expanding.

Exercises

1 The first translation of Euclid into English appeared in 1570. In the Preface John Dee pleaded the case for making such mathematical works available to those who were 'unlatined . . . and not university scholars'. He compared the situation in England with that in Italy, Germany, Spain and Portugal where there existed 'very large volumes all in their vulgar speech'. Investigate the growth of publishing in the vernacular (references such as D. E. Smith's *Rara Arithmetica* will help), distinguishing between books for the citizen and those for the scholar.

2 John Dee's *The Mathematical Praeface* (see the 1975 reprint) and John Pell's *Idea of Mathematics* (1634) (see Wallis, 1967) both attempted to provide a 'rationale' for mathematical education. Indicate briefly how you would attempt to up-date either of these. In particular how would you attempt to answer the first question posed by Pell: 'What *fruit* or *profit* ariseth from the study of Mathematics?'

3 Robert Recorde (see Easton, 1966) when preparing his English geometry *The Pathway to Knowledge* (1551) attempted to use existing or new words such as 'lozenge' and 'threelike' to replace the 'classical' terms 'rhombus' and 'equilateral'. The attempt failed; in later books Recorde changed his policy, and to this date classical terms predominate in English geometries. This is not true to the same extent in German mathematical writing. Investigate the reasons for this. The problem of either accepting alien terms or creating new ones still arises today (see UNESCO, 1975). The policy decision in Tanzania has been (like that of Recorde's) to avoid importing terms and to derive new ones from the existing vernacular: thus 'diagonal' has become 'ulalo' a word used by those constructing beds by cross-stringing to signify 'the longest (string) of all'. Argue the case for and against the creation of 'vernacular' terms as opposed to the adoption of 'international' ones.

4 D. E. Smith (1970) says of Scheubel's *Deutsche Arithmetica* (1545): '(it) is the production of a scholar rather than a man conversant with the demands of business. While Scheubel tried to write a mercantile arithmetic, the result was far removed from the needs of the common

people. It carries the work in subjects like the roots so far that the ordinary Rechenmeister could not have used it . . . Yet Scheubel, a university professor, was centuries ahead of his time in banishing the expression 'rule of three' and substituting 'rule of proportion' thus opening the way to a more mathematical and clearer understanding of this particular problem.' Discuss some of the issues raised by this example, their significance today, and measures that can be taken to alleviate difficulties.

5 Vives, the Spanish tutor of (Queen) Mary, daughter of Henry VIII, argued for a place for mathematics in the curriculum on the grounds that it provided discipline for 'flighty and restless intellects which . . . shrink from . . . the toil of a continued effort'. Yet he felt that mathematics also 'leads away from the things of life, and estranges men from a perception of what conduces to the common weal'. (Watson, 1913.) To what extent are Vives' arguments still valid?

6 Some early authors (e.g. Recorde and his German contemporary Lossius) wrote their books in the form of a dialogue between master and pupil. (Others set key theorems, definitions and even the whole work in verse, thus: 'All primes together have no common measure/Exceeding an ace which is all their treasure.') Investigate later attempts, e.g. Lakatos (1976), to revive the dialogue style.

7 Education as directed by the church (whether Christian, Jewish or Moslem) was concerned with commitment to a revealed body of knowledge. Investigate how such a view of education still permeates some national systems.

8 Investigate further the idea of 'progression' and its influence on the curricula of universities.

9 Discuss the thesis that although the fruits of 'progression' have affected the curriculum, the concept of 'progression' and the need to prepare for it have had relatively small influence.

10 What are the specific reasons why 'the SMP could have no French analogue, . . .' (see p. 16)? If you cannot answer this question now, attempt it again when you have finished reading this book.

2. **Education and the State**

During the eighteenth century various attempts were made to create educational alternatives to the Latin grammar school. The need to provide an education more suited for future merchants, industrialists, surveyors and navigators was met in America by the establishment of private academies. In their curricula mathematics found a place both for its practical value and as a mental discipline. Similar academies were established by the dissenters in England, while in Germany, Francke established a Realschule in Berlin intended to prepare boys for vocations rather than professions.

Advancing naval and military technology created a need for mathematically skilled officers and led to the establishment of institutes such as the Royal Naval (1722) and Royal Military (1741) Academies in England, the Ecole Polytechnique (1794) in France, and West Point (1802) in America.

Gradually the idea began to be accepted that education was a national responsibility and that an educated populace was a valuable asset, that education could be viewed as an investment, both personal and national. In England the task of providing popular education was for many years to be left in the hands of voluntary, religious bodies. Not so, however, in Prussia where the involvement of the state in education was such that by 1794 all universities and schools became state institutions. Even before this the *Abitur* examination had been established (1788) as an entrance qualification for university. Significantly, the *Abitur* was initially administered by individual schools, but in an attempt to increase standardisation, the ministry assumed greater control.

That France lagged behind Prussia in establishing a national system of education was not due to any lack of advocates. The case for putting education in the hands of the State was argued by a succession of writers from La Chalotais to Diderot. A minimal scheme was in fact adopted by the Convention in 1795 but it was not until the time of Napoleon that major action followed. French education was totally reorganised and all public education (with the exception of primary education) came under State control. The curriculum of the Lycées was established by law in 1809 and a highly centralised system was created: centralisation, it was argued, would promote and develop a sense of national unity. The anomaly of 'private' primary education was ended in 1833 by which time the principle that schools belong to the government rather than the people was becoming widely accepted outside of France.

In contrast to 'centralised' France, developments elsewhere proceeded in a more piecemeal fashion. Education in the USA was a state matter and states varied enormously in their prosperity, potential and practices; thus, for example, Massachusetts introduced compulsory education in 1852, Mississippi not until 66 years later. In England attempts to found a State system were initially unsuccessful, yet the compromise of State funding for voluntary bodies ironically had the effect of providing England with its sole experience of a centralised curriculum.

Gradually, however, responsibility for education shifted from the Church to the State although in certain countries, such as England and the Netherlands, the Church still had a subordinate, yet important, role to play. Moreover, in many colonies, the Church was still the principal provider of education.

Once education became a governmental interest, then what was happening elsewhere could not be neglected. In 1831 the French government commissioned a *Report on the State of Public Education in Prussia,* an initiative that was not only to be followed by a series of fact-finding missions, but which also had immediate repercussions in the USA, for the report was read and acted upon there. Henry Barnard (the first US Commissioner of Education) was soon to publish an account of his observations in *National Education in Europe* (1854), and Matthew Arnold, poet and Inspector of Schools, wrote several reports of his visits to overseas educational systems. Education had become institutionalised.

Exercises

1 In 1868 the Taunton Commission reported on secondary education in England. It found that three different types of education were needed: one for the sons of 'men with considerable incomes, or professional men, and men in business', one for the sons of 'the mercantile classes', and one for the sons of others. Mathematics occupied a markedly different place in the three corresponding curricula. The Norwood Committee in 1941 proposed a similar 'tripartite' system of education; this time, however, dependent not on parental income and standing but on the child's ability. Investigate the growth of 'two-track' and 'three-track' educational systems in various countries paying attention to the social background and the place of mathematics in each type of school. (We note, for example, that, according to J. Hitpass (1967), in West Germany elementary schools were mainly attended by children of manual and farm workers, secondary modern schools (Realschule) by those of skilled and white collar workers, and the grammar schools (Gymnasium) by children of parents with a university education and of civil servants.)

2 Dr. J. S. Howson put forward the view in 1859 that girls, because of 'their more excitable and sensitive constitutions', were ill-suited to examinations. Indeed, in England at that time, organised secondary education for girls was still in its infancy. Investigate the growth of the provisions for female education in your country and the place of mathematics in the curriculum. (Notwithstanding the good doctor's

advice, Cambridge permitted girls to enter their 'school' examinations in 1863.)

3   Reporting to the Massachusetts Board of Education in 1844, Horace Mann argued that 'the most potent reason for Prussian backwardness and incompetency is this – when the children come out from school, they have little use either for the faculties that have been developed, or for the knowledge that has been acquired. Their resources are not brought into demand; their powers are not roused and strengthened by exercise.' Is Mann's criticism still valid (outside Prussia) today? Is this state of affairs inevitable?

4   'I dare claim for the nation an education which depends only on the State, because it belongs essentially to the State; because every State has an inalienable and indefeasible right to instruct its members; because, finally, the children of the State ought to be educated by the members of the State. It is the State, it is the larger part of the nation that must be kept principally in view in education.' (La Chalotais, *Essai d'Education Nationale*, 1763.)

'That the whole or any large part of the education of the people should be in State hands, I go as far as anyone in deprecating. All that has been said on the importance of individuality of character, and diversity in opinions and modes of conduct, involves, as of the same unspeakable importance, diversity of education. A general State education is a mere contrivance for moulding people to be exactly like one another: and as the mould in which it casts them is that which pleases the predominant power in the government, whether this be a monarch, a priesthood, an aristocracy, or the majority of the existing generation; in proportion as it is efficient and successful it establishes a despotism over the mind, leading by natural tendency to one over the body.' (J. S. Mill, *On Liberty*, 1859.) Discuss.

5   'Schools have been created to maintain the social order.' (Dalin, 1978.) Discuss.

3. **The growth of professionalism**

Twenty years ago anybody was considered good enough for a schoolmaster. If a tradesman failed in business, he was thought to be learned enough for a schoolmaster; a feeble, sickly youth, who was not considered strong enough to practise any regular trade, was thought to be sufficiently qualified to undertake the duties of schoolkeeping; if a mechanic happened to get a limb fractured he would, as a matter of course, save himself from starvation by opening a school; when a man who had seen better days applied to the parish officers for out-door relief they gravely debated the question whether . . . to send him to the quarry to break stones or to confer upon him the office of parish schoolmaster. Such was the low estimate formed of the qualifications requisite for a schoolmaster. . . . But . . . a change in public opinion has been gradually taking place: the working and middle classes have been led to see the value of a sound elementary education and thereby to estimate more highly the difficulties and importance of the duties of the common schoolmaster . . . I confidently hope that the day is not distant when the force of public opinion will elevate education into the rank of a recognised science . . . based upon the nature of the being to be educated; that is to say, upon the laws which govern the development of the intellectual and moral faculties . . . laws (which) may be determined as well by observation as by psychological analysis.

This passage, written in England in the 1850s, reflects the dawn of a new era in which the teacher is viewed as a professional who must acquire not only knowledge of content, but also teaching skills and an understanding of pupils and the way in which they learn. Such pedagogical knowledge could only come as the result of special training provided in a new type of institution. In 1826 Prussia instituted a state qualifying examination for teachers in the vernacular elementary schools and by 1840 had 46 teachers' seminaries. This latter year saw the opening of the Battersea Training School in England, a private venture (in which Tate, the author quoted above, was to tutor), one year after that of its US counterpart, the State Normal School at Lexington, Mass.

Significantly, within a few years questions were raised in England as to the balance between professional and academic studies that should be sought in these new institutions: a problem still awaiting resolution.

When elaborating his views on a science of education, Tate was much influenced by Rousseau, Pestalozzi and the English philosopher Locke. From Locke he absorbed the ideas of 'faculty psychology', that every individual possesses 'intellectual and moral faculties' which 'are cultivated by being *properly* exercised on appropriate subjects'. Here then was a basis on which a curriculum could be planned and a subject's position in it justified: thus,

> *Mental arithmetic* cultivates the memory and the powers of conception and reasoning. It also especially fosters the habit of promptitude, presence of mind, and mental activity.
> *Arithmetic* cultivates the reasoning powers and induces habits of exactness and order.
> *Mathematics and Natural Philosophy* cultivate the reasoning powers chiefly in relation to the acquisition of necessary truths; they also cultivate habits of abstraction. (Tate, 1857, p. 249.)

Faculty psychology had, however, already been challenged in Prussia by Herbart (1776–1841), an educator who was to have considerable influence on later developments in the USA and Britain. (The National Society for the Study of Education grew out of the National Herbart Society.) Herbart argued that one should not consider separate faculties, but rather degrees of activity and that, above all, one should seek to cultivate interest: 'to be wearisome is the cardinal sin of instruction'. In the ordinary course of life the child would acquire interests connected with nature and the physical world and also through social intercourse: one should build on from there, new matter being woven into knowledge already gained with a growing unification of interests in the intellectual sphere.

> What is the bearing of all this on curriculum-building? The formalist frames his curriculum with the object of developing certain faculties. The Herbartian allows the mental content to grow by laying hold of ideas that will hook on to the old . . . And the old will include not only old knowledge but old experience.

The author, Charles Godfrey, was an influential English mathematical educator. Here, writing in 1911, he reiterates Tate's view that the curriculum must be planned on educational principles. That those principles are constantly evolving serves only to demonstrate the difficulty of the educational psychologist's task: it does not detract from the underlying argument.

Godfrey did in fact strive hard and with considerable effect to change the mathematics curriculum in English schools. Yet he was unable to bring about changes through his own efforts. The latter part of the nineteenth century had seen the growth of a powerful examination system in England in which the universities wielded enormous power. One man alone could not battle successfully against those whose decisions so constrained the school curriculum. In an institutionalised system significant progress could only be achieved by concerted action: a fact which in England led to the formation in 1871 of the Association for the Improvement of Geometrical Teaching. This association comprised schoolteachers, fortified with university lecturers, who sought to persuade examination bodies to approve other approaches to geometry than that of Euclid. Their fight was not immediately successful, indeed it was not until 1903 that Oxford and Cambridge finally agreed that Euclid's rule should be brought to an end. (Here it is essential to point out that when Dieudonné demanded in 1959 that 'Euclid must go' (see p. 102) he was speaking of the 'spirit' of Euclid; the AIGT fought, however, against the 'letter' of Euclid, the use of his *Elements*.) By that time the AIGT had broadened its interests to other areas of mathematics and accordingly changed its name to 'The Mathematical Association'. In 1894 it published the first number of the *Mathematical Gazette,* a journal containing articles aimed at improving mathematical teaching. It was not the first such journal; *Mathesis* commenced publication in 1871 and *L'Enseignement Mathématique* in 1893.

Following the example of the AIGT, professional associations came to be formed in many countries, for example, the Association of Teachers of Mathematics in the Middle States and Maryland in 1903. The latter's journal *The Mathematics Teacher* was taken over by the National Council of Teachers of Mathematics after its foundation in 1920.

The professional associations were responsible for great improvements in mathematical education. They encouraged and provided a means for the exchange of new ideas. However, their influence tended to be limited to the secondary school sector which at the beginning of this century was relatively small and almost entirely academically biased. The rapid expansion of secondary education and the accompanying need to rethink the purposes of mathematical education posed problems that the associations were rarely able to meet. As agents of change they were no longer adequate: as we shall see, new agencies had to be provided.

Yet the association can still have a leading role to play in curriculum development. In countries with centralised systems, such as Poland and France, the association is a powerful means for exerting influence on the education ministry. In developing countries the young, small and vigorous association has often been in the forefront of development work, as in Ghana. Where an older association exists there is always the possibility of a splinter group of reformers detaching itself and forming a new association. This occurred in Britain in the 1950s and led to the founding of the Association of Teachers of Mathematics (ATM). Through associations' journals and reports, and at their annual conferences reformers can still make their work known to a large audience. Backing and support can be provided for small groups of innovators.

The opening of the twentieth century witnessed changes in mathematics teaching in countries other than Britain. In the USA, E. H. Moore, a leading mathematician, urged the cause of reform, D. E. Smith attempted to establish the foundations of a discipline of mathematical education, and G. B. Halsted produced a school geometry based, not on

Euclid but on Hilbert. In France a government decree of July 1905 invited 'teachers to follow a method entirely new in geometry': that the 'entirely new' was basically a book published 30 years earlier need neither worry nor surprise us! In Germany, Klein was giving the lectures which were to be published later as *Elementary Mathematics from an Advanced Standpoint.*

The need to exchange information and experience was clear. In 1905 Smith, writing in *L'Enseignement Mathématique,* floated the idea of an International Commission, an idea which was adopted at the 1908 meeting of the International Congress of Mathematicians.

This commission, known as the 'Commission Internationale de l'Enseignement Mathématique' (CIEM), had Klein as its President and was responsible for mounting a number of meetings and enquiries. Thus, for example, at a meeting in Milan in 1911 it directed attention to three specific questions – 'What Mathematics Should be Taught to Those Students Studying the Physical and Natural Sciences?' 'What is the Place of Rigour in Mathematics Teaching?' and 'How Can the Teaching of the Different Branches of Mathematics Best be Integrated?'; three questions that are equally relevant today.

The main work of the CIEM was, however, the preparation of a vast report on teaching practices in member countries. Each country prepared a national report often running to many volumes (eleven in the case of the USA!). Certainly nothing on the same scale had been attempted before, or has been attempted since.

Unfortunately, the CIEM was to be one of the many victims of the Great War of 1914–18. Not until the 1960s was international cooperation again to reach such a level. By that time curriculum development enjoyed a new status. No longer could the curriculum be left to evolve in a natural way. The rate of change had quickened to such an extent, and the educational systems expanded so much, that agencies and bodies had had to be established specifically to promote and implement change: curriculum development had itself become institutionalised.

Exercises

1 Compare an early volume of the *Mathematical Gazette* (or your country's equivalent) with a recent one. What differences in aims, content and approach can be discerned?

2 What role does the subject association play in your country? Discuss the effectiveness with which it meets its aims, and their appositeness, and the constraints which beset it.

3 Write an essay on one of the topics debated at the 1911 meeting of the CIEM (see above).

4 'If we send forth the teacher to the discharge of his lowly but momentous duties with, in most cases, only a moderate range of attainment, let us provide that he have acquired such a readiness on all that concerns the art of teaching as will render his knowledge at once available. Nothing like this has yet been satisfactorily realised by any of our Training Colleges.' (Minutes of the Committee of Council, 1847–8.) Discuss. Investigate the history in Colleges of Education of the 'General Education *v* Professional Training' controversy.

5 'It is characteristic of *the form* of instruction adopted here that it is chiefly *oral*. The students do not acquire their knowledge from books but directly from teachers. . . .

'It is a further recommendation of the method of *oral* instruction . . . that it is the method of instruction which the students may hereafter apply with the greatest advantage in their own schools. It is by *oral* instruction alone that poor ignorant children can be led on, for the first time, to think, to reason, and to understand. . . .

'It is, moreover, a consideration . . . that oral instruction is a means by which great *moral ascendancy* is given to the teacher over the minds of his scholars . . . It is a quiet demonstration of intellectual ascendancy, from which there is no escape . . ..

'Notwithstanding . . . it is the defect of oral instruction that it does not make *students;* it neither gives the habit of self-instruction, nor the taste for it. The difficulties opposed to the acquisition of knowledge are in a lecture smoothed away; the mind accustoms itself to lean upon the lecturer . . .'

Discuss the points raised in these passages taken from the Reverend Henry Moseley's official report on his visit to Battersea Training School in February, 1846.

6 'Examinations directed, as the paper examinations of the numerous examining boards now flourishing are directed, to finding out what the pupil *knows,* have the effect of concentrating the teacher's effort upon the least important part of his function.' (Mark Pattison, 1880.) Is this criticism still valid, and if so what steps can be taken to remedy matters? What are the more 'important' parts of the teacher's function? What are the consequences for mathematics teaching?

7 In 1838 the Committee of the Central Society of Education offered a prize of £105 for the best essay on 'The Expediency and Means of elevating the Profession of the Educator in Public Estimation'. The authors of this book cannot offer a comparable financial inducement, but invite the reader to write an essay on that theme. The 1838 winner (see Lalor *et al.*, 1839) considered the proposal 'to connect with each of

the Universities a professor of education, whose duty it would be to deliver lectures', but rejected this suggestion as unlikely to produce much of benefit. Professors of Education may, therefore, wish to write on the theme 'Why Lalor was wrong'.

4.   **The origins of curriculum theory and reform**

We have seen how in the nineteenth century it was beginning to be accepted that the school curriculum should be built upon pedagogical principles which accorded due weight to psychological theories. Yet the available theories of education and pedagogy were too general to form the basis for planning a curriculum; whilst, on the other hand, those principles which governed the construction of the syllabus (Lehrplan, programme) were too narrow. An amalgam was needed. Moreover, rigour had to be brought to a field dominated, to that time, by philosophical arguments and justifications, and by pedagogical and psychological intuitions. The need to treat curricular problems scientifically arose for several reasons.

The legitimacy of educational practice is always subject to question when changes are being proposed or opposed. No greater change, however, can be imagined than that process of democratisation that began to affect schools about the beginning of this century; opening schools (particularly those supplying secondary education) to a greater section of the population and necessitating changes in organisation, aims and content. The content was no longer that for an élite and because of its greater applicability, it demanded more consideration. Philosophically-based ideals were no longer sufficient to justify the curriculum; demands on, and by, society had to be considered to a far greater extent.

Freer access to schools resulted in a development and extension of examination systems. This, in turn, led to a more critical review of the criteria underlying curriculum construction, and to the evaluation of intentions and their realisation: concerns not only of those directly concerned with the examination process but also of those affected by it. Other organisational changes (for example, the opportunities presented to students to select their own courses) raised further questions concerning value and utility. In short, the democratisation of the educational system made it more necessary to justify curricular decisions and actions, and to provide consistent and logical arguments on their behalf.

The ways in which countries have responded to this dynamic development in society and science have differed and have been greatly affected by traditions. The first approach to the problem through empirical research, and the first attempt at establishing what might be called a curriculum theory, was undertaken early in this century by John Dewey in the USA. We shall briefly survey Dewey's theory and its consequences in chapter 5, but it is perhaps useful first to consider how earlier approaches to curriculum reform had been conceived in European countries.

In fact, two very different conceptions had emerged, one on the continent and one in England, and their consequences can still be felt today.

The French revolution sought to abolish the clergy's domination of education and to establish a school system for the new society (cf. La Chalotais, p. 23). Important means to this end were the secularisation of schools and the attempted dissemination of rationalism through the introduction of modern science into the university curriculum. The rise of the French nation in the Napoleonic era was partly attributed to the efficiency of its army commanded by officers trained at the Ecole Polytechnique. Prussia, which had suffered much from that army's effectiveness, was duly impressed. French ideas of mathematically and scientifically oriented university studies were adopted by Prussian universities, and, later, the curriculum of the Gymnasium was questioned: no longer did the classical programme seem appropriate. A comprehensive educational theory emerged (which itself was to influence the French Lycée curriculum after 1830). The Humboldt reforms of 1810–1819, based on the idealist view of the individual, developed the concept of 'Bildung'. In this, 'Bildung' (education) became the paramount goal of all human intellectual and mental development. 'Bildung' was to be interpreted in two ways: as the totality of knowledge and judgement, and as the process of education. Correspondingly, science had a dual role in this concept: it provided essential subjects of 'Bildung', and, through scientific methods and procedures, it represented the learning process at its most advanced level. Science, therefore, not only contributed content, but acted as a guide as well. Thus the concept of 'Bildung' established a unity of science and education, of research and instruction. Moreover, 'Bildung' constituted a higher-level union of the heterogeneous subjects of science and so countered the dangers of compartmentalisation, as a result of the admission and growth of new subjects in education. The concept of 'Bildung' also served to mediate between the social and individual demands in education. The theory was itself individual-centred, but since the subordination of individual to social demands was regarded as one of the highest virtues, the claims of society were automatically safeguarded.

This conciliatory potential peculiar to the theory of 'Bildung' made it an extraordinarily workable instrument of educational policy. Yet its generality and lack of precision meant that it could also be invoked by, say, those who, whilst pretending to further the education of individuals, actually discriminated against this, and who practised drill in the name of free personality development, or dull grind in the name of science.

Another crucial flaw of the 'Bildung' theory was that it led to a belief in an ordained method of teaching associated in a natural manner with each subject. Thus psychological advances and theories of teaching and learning could not be readily assimilated. Just as the original universality of the concept faded away during the nineteenth century, or could no longer be maintained, so the concept of 'Bildung' shrank to a mere theory of the syllabus, which, incidentally, was conceived as the *maximum* to be taught, rather than the minimum. Many of the present problems of German curriculum development derive from the resulting overemphasis on syllabus content and the difficulty in assigning equal value to pedagogy as a professional component in teacher training.

A quite contrasting development took place in England, where, as the quotation from Mill (p. 23) makes clear, there was strong opposition to the education of children by, and for, the State. A different view of personality formation moulded the goals of school education. Not the scholar, but the gentleman, was the ideal of education: a harmonious development of all mental and physical powers was sought; the school providing a balance of individuality and social convention. In consequence, there was a tendency to pay more attention to methods of teaching and to pedagogy in general. This was illustrated in many of the reform movements, for example, in the work of the Hills and the Mayos (see, e.g. Stewart & McCann, 1967). Moreover, changes in the curriculum often meant not so much a reshuffling of the academic content (for this proved difficult to achieve), but a widening in ideas of what constituted a curriculum. Herbert Spencer was to argue that children should be trained 'not only to fit them mentally for the struggle before them, but also to make them physically fit to bear its excessive wear and tear', an argument that manifested itself in a cult of organised games and the production of 'muscular Christians'.

Possibly the most significant difference between these reforms and their continental counterparts arose, however, from the peculiarities of

The English school organisation. Not until the 1860s did the State exercise any serious control of elementary education. Then 'account-ability' raised its head and, under the 'payment by results' scheme, grants made by the State to elementary schools (which were still all in voluntary hands) were linked with the 'satisfactory' performance by children in an examination conducted by the inspectorate.

Not until the twentieth century was the State officially involved in secondary education. Thus, although governmental commissions in the mid-nineteenth century pressed the claims of a modern secondary-school curriculum in which mathematics would occupy a central position, there was no way in which changes could be enforced upon schools. Yet another result of this lack of involvement was that before 1900 elementary and secondary education developed in nearly total separation from one another.

Moreover, secondary education was dominated by the grammar schools, and the high esteem of this type of school made it difficult for other forms of secondary education to take root. When the opportunity to initiate change arose in 1902 it was not taken. 'The new regulations were based wholly on the tradition of the Grammar Schools and the Public Schools . . . little or nothing (was done) to foster the development of secondary schools . . . designed to meet the needs of boys and girls who desired to enter industry and commerce at the age of 16.' (Spens Report, 1938: note that this refers to the *able* and not the average pupil, who at that time left school at 14.)

The growth of State involvement in English education was not accompanied by any strong movement towards the centralisation of education. Administration was left in the hands of local education authorities (originally, over 300 in number), and each school was, in theory, in control of its own curriculum. A strong 'independent' educational system remained in existence.

The absence of a central administrative authority meant that new ideas and concepts could not be readily implemented throughout the educational system or given authoritative backing. As a result, most attempts at reform were limited in size and did not survive their originators. In consequence, there was little opportunity (or incentive) to develop a comprehensive theory of the curriculum. Little continuity existed in reform beyond a common tradition, yet this, and in particular the close involvement and participation of teachers, was to be of considerable significance and contrasted strongly with the position on the continent.

Exercises

1 'Humboldt's achievements were important for (England) because they strongly influenced the views of the Clarendon and Taunton Commissions and helped to form the ideas about the nature of a liberal culture which were expressed by Matthew Arnold.' (Curtis & Boultwood, 1965.) Investigate.

2 'The cry for (education) among the lower orders is because they think that, when once they have got it, they must become upper orders . . . very sternly I say to you . . . that a man had better not know how to read and write, than receive education on such terms.' (Ruskin, *Time and Tide,* 1868, 2nd edn.) How have society's views on education changed in the century since Ruskin wrote this?

3 'The primary business of the school is to train children in co-operative and mutually helpful living.' (Dewey, *The Elementary School Record,* 1900.) To what extent is this a peculiar, United States view of the aims of education? Would it receive support in your country today? Assuming that this is the primary aim of schooling, what contribution can mathematics make to its attainment?

4 'A man is not educated, in any sense whatsoever, because he can read Latin or English, or can behave himself in a drawing-room; but he is only educated, if he is happy, busy, beneficient, and effective in the world: millions of peasants are, therefore, at this moment, better educated than most who would call themselves gentlemen.' (Ruskin, *Stones of Venice,* 1851.) To what extent is this an eccentric, English view of education? If one accepts Ruskin's argument, what are the consequences for mathematics teaching in our schools?

5 Two important landmarks in the movement which led to secondary education for all in Britain and the USA were the Kalamazoo case (State of Michigan, 1874) and the Cockerton Judgement (Queen's Bench, 1901). Compare and contrast the preliminaries to the two cases, the two findings and their consequences.

6 'The genesis of knowledge in the individual, must follow the same course as the genesis of knowledge in the race.' (Herbert Spencer, *Intellectual Education,* 1854.) This principle of curriculum design has been reiterated by Poincaré, Polya, Thom and others, frequently under the guise of Haeckel's Law: 'ontogeny recapitulates phylogeny'. (Spencer, we see, has prior claims to Haeckel (1834–1891) but he himself attributes the notion to Auguste Comte (1798–1857).) Discuss the validity of this principle and its consequences for the design of mathematical curricula.

7 'We must not expect a child to "invent geometry", as Mr Spencer suggests and Pestalozzi also desired. Pascals are rare in the world, and

few could follow in the steps of Euclid.' (Compayré, 1908.) Investigate the history of 'discovery' methods. How would you reply to Compayré (or Spencer)?

8 'Today Great Britain and the United States are the classic lands of open-air games . . .. This violent education of the body . . . develops energy of character; but its first effect, naturally, is that of developing physical qualities. It is with good reason that M. Maurice de Fleury, a Frenchman, attributes partly to this the difference in build and manner of the two races: the Englishman tends to become tall and graceful; the Frenchman, short, thickset, and effeminate. (Compayré, 1908.) Comment!

3

Case Studies of Curriculum Development

1.   **Three projects**
      Before embarking on an analysis of how curriculum develop-
ment in school mathematics has been managed, and how this management reflects various theories of curriculum change, the reader should
be aware of the variety of forms that curriculum development can take.
In this chapter we present case studies of three curriculum development
projects in mathematics: The Fife Mathematics Project of Scotland, the
Secondary School Mathematics Curriculum Improvement Study of the
USA, and the Individualised Mathematics Instruction Project of
Sweden. These three projects were chosen not because they are either
typical or exemplary, but because together they serve to illustrate,
about as well as three projects can, the variety of approaches that have
been taken to the problem of managing mathematics curriculum development.
      Each project is sketched in turn, and then together they are discussed
as they relate to certain issues of management. Reporting on the
Allerton Park conference (see p. 88) on styles of curriculum development, Stuart Maclure (1972) noted that to see contrasts in 'style' and in
underlying values, one should look at issues that divide opinion rather
than issues on which there is consensus. Maclure identified three key
issues that divided opinion at the conference: (1) the contrast between
centralised and decentralised systems, (2) the impact of curriculum
development on the role of the teacher, and (3) the relation between the
centre and the periphery. We have used these same three issues to
illustrate some of the ways curriculum development projects can differ.
The three countries from which the case-study projects come vary in the
degree of centralisation of their educational systems. They also differ in

their national beliefs, or 'myths', regarding the teacher's role with respect to curriculum development. One can see from the case studies how these differences are reflected in the projects. One can also see that the locus of curriculum development can be either the centre or the periphery of the system (Schon, 1971) and that dissemination, if it occurs, can occur in various directions.

The case studies are intended to provide an introduction to the discussions in subsequent chapters. The brief descriptions of the projects cannot really do them justice, nor can the three dimensions of differences capture the manifold ways they differ. But this chapter should at least suggest some of the variation one finds within curriculum development. In subsequent chapters we shall offer a more elaborate analysis of this variation.

1.1. *The Fife Mathematics Project*

The Fife Mathematics Project began as a response to the movement towards comprehensive education in the secondary schools of Fife (Crawford, 1975). Many mathematics teachers were concerned about how to organise instruction in mixed-ability classes. In May 1969, Geoffrey Giles, Lecturer in Mathematics in the Department of Education at Stirling University, was asked by the Organiser of Secondary Education in Fife to speak to mathematics teachers in the county. Giles proposed the individualisation of some of the mathematics instruction in the first two years of secondary school by means of a workcard system that would allow pupils to work alone or in groups. After some discussion with school authorities, Giles agreed to conduct a pilot experiment to begin in February 1970 in two high schools. In September 1969, he met with mathematics teachers in the two schools to discuss the proposed experiment. One of the schools later withdrew because of a staff shortage. The staff of the remaining school, who had been instructed by their Principal (Head) to participate in the project, carried on (Crawford, 1975). In Giles' original formulation, the purpose of the experiment was to try out the workcard approach in a relatively free classroom situation in which the pupils could choose activities and co-workers. Effects and difficulties of the approach would be explored, and it would be demonstrated for others to see and judge.

The experiment proper began in February with two classes of first-year pupils who would not actually begin their secondary programme until the following August. Giles taught an introductory lesson to the pupils. He returned periodically to monitor the progress of the classes. The instructional material, which had to be shared by the two classes, was supplied by the project.

By May 1970, Giles had reformulated the aims of the work as follows.

Educational aims of the experimental work

1. To stimulate the learning and understanding of each child through his active involvement in the work;
2. to foster a sense of responsibility by leaving to the pupil many decisions about the use of his time;
3. to encourage self-reliance by providing a range of reference material to which the pupil has immediate access;
4. to develop self-confidence in the pupil by allowing him to experience a sense of achievement after individual or co-operative effort.

Mathematical aims

1. To increase the pupil's familiarity with the basic concepts of number and shape, and with the derived concepts of mappings and relations;
2. to develop his ability to think logically and to generalise explicitly from simple patterns;
3. to give him the experience of personal exploration in a range of carefully-chosen situations that will help him to appreciate the meaning and significance of the more formal mathematics that will come later. (Crawford, 1975, p. 10.)

Although the mathematical aims continued to be important, Giles found himself giving increasing emphasis to the educational aims. After a year of experimental work, the interest and enthusiasm of the pupils began to diminish. Class teaching was gradually introduced to ease the transition to regular work when the experimental classes joined the other classes for the second year of secondary school.

In June 1971, Giles conducted a weekend conference at Stirling for 20 principal mathematics teachers from other schools in Fife who were considering participation in a revised and expanded project. The major revisions consisted in combining workcards on the same topic into booklets and in using a flexible combination of class teaching and individualised instruction; both revisions were intended to give more direction to the pupils' learning without unduly restricting their freedom to choose learning activities.

Over the summer, Giles revised the workcards and organised them into booklets that were printed by the Fife Education Authority. In the autumn of 1971, 28 classes in eight schools began working with the revised project materials. Officials in the Education Authority decided during that school year that the project should be offered to all Fife secondary schools during the following year, with costs met by the Authority. In 1971–2, then, the project was expanded to the 19 available secondary schools in Fife (the remaining two were pilot schools in a separate project).

At this point, a separation occurred: the 'Fife Mathematics Project' came to refer to the method of working in the classroom that the teachers in Fife had developed, quite apart from the materials developed by Giles. Giles stepped out of the project, revising the booklets containing the workcards and adding testing materials. Mathematics Workcard Booklets were made available to a wider audience through the University of Stirling. The Fife Mathematics Project continued as a project of the Fife Education Authority, with teachers producing their own materials and adapting others. Some schools have for various reasons decided not to continue with the project, but most schools in the county still use the project as a supplement to the syllabus and more traditional methods. Several schools outside Fife have used the project's approach and materials.

1.2. *The Secondary School Mathematics Curriculum Improvement Study (SSMCIS)*
Most recommendations for reforming the US secondary school mathematics curriculum during the 1950s and early 1960s assumed that arithmetic, algebra, and geometry would continue to be taught in separate year-long courses. Revisions were suggested within, but not across, course boundaries. European countries did not have this sharp separation in course content, and their reforms tended to strengthen even further the unity of the school mathematics curriculum. The Secondary School Mathematics Curriculum Improvement Study (SSMCIS) was begun by Howard F. Fehr, of Teachers College, Columbia University, as an experiment in developing a unified six-year curriculum. Restrictions of traditional content or sequence were to be discarded in favour of a syllabus that would 'encompass all the mathematics that is now considered essential through a first year university program, specifically the fundamental structures of number systems and of algebra, linear algebra, probability and its applications, mathematics related to computers, and the calculus (analysis)' (Fehr, 1966, p. 533). A meeting of consultants (mathematicians and educators)

was held in November 1965 to outline 'a procedure for developing a total program, to be followed by syllabus conferences, writing of experimental textbooks, education of the classroom teachers, pilot class teaching, and evaluation of outcomes'. (SSMCIS, 1969, p. 2.) A grant was obtained from the US Office of Education to support SSMICS, and at a 20-day meeting in June 1966, a group of eighteen mathematicians and educators from Europe and the United States outlined the 'scope and sequence' of the entire six-year programme and made specific recommendations for the content of the first course. A team of eight mathematics educators, all of whom had taught in secondary schools, took the recommendations and wrote a preliminary version of the first course during the remainder of the summer. At the same time, special courses in the mathematics and pedagogy of the proposed materials were given to the twenty teachers who would teach the experimental course to ten seventh grade* classes (two teachers to each class) during the following year. In June 1967, another syllabus conference was held to consider revisions of the material already written and to make specific recommendations about the second course (for the eighth grade). The conference was followed by an eight-week writing workshop to revise the first course and prepare the second and by six-week teacher education courses related to each of the two experimental courses. During the 1967–8 school year, the preliminary version of Course 2 and the revised version of Course 1 were tried out in the pilot schools. This procedure continued each summer and the following school year until 1973, when revised versions of all six courses had been completed (and the first five revised courses had each received an additional year's trial and final revision). Grants from the US Office of Education supported the development of the Courses 1 to 3; grants from the National Science Foundation supported the development of the Courses 4 to 6 as well as the dissemination of project materials and summer and academic year institutes to train teachers to use the programme. The experimental teaching of the six courses was done in five school systems in the metropolitan New York area, but as revised courses became available the programme spread to more schools. Reporting on the scheme in January 1974, Fehr wrote that 'today in the United States from Long Island to Los Angeles there are 25,000 students in Courses 1 through 6'. (Fehr, 1974, pp. 31–32.)

The SSMCIS programme was designed for the most capable students:

---

* The 'grade' system is explained in Appendix 1, p. 267.

'the top 15 to 20 percent of secondary school students'. (Fehr, 1974, p. 33.) The criteria for selecting students are left vague in SSMCIS publications (phrases such as 'upper 15% of mathematical ability' and 'upper 15% of cognitive ability' are used, and the percentage is sometimes given as 10 rather than 15 or 20). Each school system must arrange its own selection procedures, although most have depended on some combination of sixth grade teacher's recommendations and intelligence test scores. The programme is so different from most US secondary mathematics curricula that it is almost impossible for a student to enter it after Course 1. Students do drop out, however, and a school system usually needs to start at least two classes in Course 1 to make sure that enough students remain in the programme until twelfth grade so that there can be one class in Course 6.

Most US school systems are organised so that students enter high school in either the ninth or tenth grade. Consequently, school systems must arrange for the programme to continue as students progress from middle or junior high school to high school. The problem is compounded when, as often happens, several schools feed into the same high school. During the development of the SSMCIS courses, the special demands on the teachers in the pilot schools essentially required that they moved along from grade to grade with their students: other teachers were not prepared to teach the courses. Students in the experimental classes, therefore, had the same teacher (in most cases, two teachers) for six years, whereas they ordinarily would have had a different mathematics teacher each year. Many teachers in the experimental programme used their experience and additional training to move into new positions on high school faculties, and at least three of the original twenty teachers moved into college and university positions.

The teachers in the experimental programme recorded their experiences in trying out the materials. At intervals during the school year they reported to the SSMCIS staff their criticisms and suggestions for revising the material. Some of them also assisted in preparing teachers' commentaries for the courses and in conducting workshops, leadership conferences, and in-service training courses for other teachers.

The project made extensive use of European mathematicians as consultants, especially at the June conferences. The principal American consultants were Marshall H. Stone, Albert W. Tucker, Edgar R. Lorch, and Meyer Jordan. Other consultants were brought in from time to time: one list contains 39 names (SSMCIS, 1973, p. 23). Graduate students in the Department of Mathematical Education at Teachers

College served on the SSMCIS staff each year, helping Fehr with the administration of the project and serving as writers, consultants, teachers, and evaluators.

Courses 1, 2, and 3 are in the public domain for general use. Teachers College Press distributes paperback revisions of the revised SSMCIS Courses 1 to 6, and Addison–Wesley Publishing Company distributes a hard-cover adaptation by Fehr and others of Courses 1 to 4 that is written to be used with students of a wider range of ability. The Bureau of Mathematics Education of the State of New York is developing an integrated, unified mathematics curriculum for grades 9, 10, and 11 that attempts to present some of the SSMCIS topics to students of average ability.

1.3.   *The Individualised Mathematics Instruction (IMU) project*

In the autumn of 1963, the National Board of Education of Sweden commissioned Curt Öreberg, a teacher from Älghult/Kronoberg County, to direct a study to compare the effects of completely individualised and 'conventional' mathematics instruction in grades 7 and 8 (the first two years of secondary school). Individualisation was being advocated as a means of improving instruction in the newly introduced nine-year comprehensive school, and there was additional pressure to revise the content of school mathematics. Öreberg had been experimenting since 1959 with various ways of individualising mathematics instruction; he had been using some existing correspondence course material for instruction. The results of the study led the Board to fund the IMU (individualiserad **mathematikundervisning**) project in the autumn of 1964. Teschner (1973) describes the management of the IMU project as follows:

> The project was carried out at the Department of Educational and Psychological Research of the Malmö School of Education. The Hermods Publishing Company of Malmö agreed to cooperate in the production and future distribution of the material. Öreberg remained project leader while a scientific leader was made responsible for the scientific supervision of the project. Later, in 1969, it was thought advisable to make a more definite division of responsibility in the project management. Öreberg was made responsible for material production and for the activities connected with the practical application of the methods-materials-systems approach in the schools, while Lars M. Jivén was made responsible for data collection and evaluation.

A working group composed of an education scientist, the author group (a teacher, a mathematician and an expert in teaching material of the publishing company participating in the project) and three lecturers in the methodology of mathematics teaching carried out the analysis of objectives, on which the construction of the first experimental version of the IMU material was based. The objectives remained unchanged during the subsequent work, but there were very considerable changes in the structural and methodological design of the material. As the material was revised in a total of five versions, possible approaches in the direction of the desired individualisation were repeatedly modified in the light of performance and attitude data obtained from pupils and teachers, in each case after the testing of the corresponding preliminary versions (p. 390).

The project developed a completely new mathematics programme for grades 7, 8, and 9. Two hundred and eight separate instructional objectives in seventeen areas of mathematics were identified, and course material was developed for each objective. In the fifth version, each school year's course is divided into three modules, each composed of ten workbooks. The pupil works through the material on his own, using tests to guide his progress. Film strips, tapes, laboratory material, and a special workbook for group and formal class teaching are part of the instructional system.*

As the IMU project developed, its objectives became somewhat separate from the objectives of testing a novel approach to individualisation (as also occurred in the Fife Mathematics Project). The objectives of the IMU project were

constructing and testing self-instructional teaching material for mathematics teaching;
developing and testing suitable teaching methods for the application of this material;
investigating the grouping of the pupils and the utilisation of the teachers with a view to maximum effect of materials and methods;
measuring the effects of absolutely individualizing instruction with the help of the material constructed (Teschner, 1973, p. 368).

The objectives of the 'methods–materials–systems' approach were

making use of the opportunity afforded in the Curriculum and Study plan for selecting the subject matter according to the aptitudes and interests of the pupil;

---

* See Teschner, 1973, pp. 369–71; Crawford, 1975, pp. 134–6; or Taylor, 1972 p. 176–9; for further details of the IMU teaching material.

adapting the instruction individually to the pupil, i.e. choosing
explanations, illustrative examples, illustrative material and methods
of finding solutions according to the pupil's learning aptitudes;
selecting the difficulty and number of the exercises assigned to the
pupil according to his aptitudes;
enabling each pupil to work at his own pace (Teschner, 1973, p. 369).

The IMU project conducted 'field experiments' on the effects of individualised instruction and flexible grouping of pupils and teacher teams from 1964 to 1969. Production of the first version of the material was begun during the 1965–6 school year at one school. Nationwide field experiments were conducted for the second version from 1967 to 1969 and for the third version from 1968 to 1971.* By 1971, about 12 000 Swedish pupils were using the third version of the IMU materials. In 1972, IMU experimentation was concluded when funds were cut back; three versions had been field tested and the fifth version had been produced.

International interest in the IMU material has been high: Norway began production of a version in 1968, the UK and Ireland began experiments with translated IMU material in 1969, and Yugoslavia and West Germany have considered adaptations of IMU materials. Aspects of the IMU system, apart from the materials, have been of interest to educators in many countries.

Within Sweden, the National Board of Education did not wait for the results of experimentation to recommend regular introduction of the IMU system nationwide. In November 1968, the Board stated:

Even if the final scientific efficiency surveys have not yet been
completed, the Board has nevertheless found the data already
available so overwhelmingly positive that the application of the
methods–materials–systems approach should now be allowed in the
schools (Teschner, 1973, p. 403).

This recommendation was made without consulting the IMU researchers, and they reacted against what they saw as premature and compulsory introduction of the system into the schools. Reactions from the researchers and from the Malmö Board of Education led to a policy whereby the IMU system can be introduced only if the local school administration requests it after consultation with the teachers who will be involved. The IMU author group subsequently developed on its own a similar, but much simplified, instructional system (*Hej mathematics*)

* See Larsson, 1973, for details of the field experiments.

that is available commercially; it had about three quarters of the market by 1971 (Teschner, 1973, p. 405).

2. Contrasts between the three projects
2.1. *Centralised and decentralised systems*

The three projects sketched above come from three different countries, and each is to some extent a reflection of that country's social, political, and educational systems. The Scottish Education Department has sponsored curriculum development projects, but the tradition of local control of education and school autonomy in Scotland makes it possible for a project to emerge from the grass roots under the sponsorship of the local education authority. In the case of the Fife Mathematics Project, developments at a national level (the move to comprehensive schools) and the response of the Organiser of Secondary Education to concerns expressed by mathematics teachers triggered the project, but it would not have survived and flourished without the efforts of Geoffrey Giles and the Fife mathematics teachers who worked with him. At the same time, the Fife Education Authority supplied essential financial support. The project was a local response to a national problem. The absence of strong centralised control over the curriculum of the early secondary-school years permitted the project to be flexible in its goals and procedures. The availability of funds and support at the county level stimulated dissemination of the project to county schools, but could do little to aid dissemination outside the county.

The US has not traditionally subscribed to the myth of the teacher as determiner of the curriculum; instead, it has held to the myth that the local community is the determiner of the curriculum. As Ovsiew (1973) observes, recent curriculum development projects in the US have had a rather different premise:

> that individual differences were within a standard range so that a curriculum could be built anywhere to suit all or part of that range. Which is to say that the differences along school districts were irrelevant; only the differences among children mattered in curriculum construction. And if that were true, it was worth spending millions of dollars, using the very finest scholars, and paying for research, field test and development to create curricula that could be used in thousands of school districts, rather than spend hundreds of dollars using teachers to make curricula to be used in one school district (p. 530).

The SSMCIS project was not funded as generously as some other American mathematics curriculum development projects have been, but it is in the American vein in its attempt to devise a curriculum that could be used in thousands of school districts. Funding by national agencies guaranteed nationwide visibility and some support for nation-wide dissemination activities. School districts used for the experimental teaching showed strong support for the project while it was under way, but commitment from state educational authorities was difficult to obtain. The SSMCIS programme makes substantial demands on a school district: it requires special administrative arrangements and qualified teachers. A more centralised system might have provided greater incentives for districts to adopt the scheme.

The IMU project, like the Fife project, began at the grass roots, in a single school, but was taken under the wing of a regional educational authority. The relative centralisation of Swedish education is reflected, however, in the pressure to implement IMU nationwide (in fact, before its developers were ready). The Swedish myth is that the government should set the framework for curriculum development. 'At the local level most professionals, including the teachers, do *welcome* central initiative in educational innovation. This is accepted to such a degree that it would be unthinkable, at the local level, to initiate innovations which do not conform with the broader aims set out at the central level.' (Dalin, 1973, p. 25.) In a centralised system like that of Sweden, curriculum innovations can be picked up and implemented rapidly, but only if they are in harmony with the 'broader aims set out at the central level'. Would a Swedish SSMCIS project have received central support? Would a Swedish Fife Mathematics Project have received central support?

2.2. *The role of the teacher*

In the Fife Mathematics Project teachers' concerns about indi-vidualisation helped initiate the project, and their various attempts to use the workcards helped shift the project's focus from content to methodology. Teachers were not so much creators of materials as they were collaborators in converting the methodology from complete indi-vidualisation to a flexible combination of class teaching and individual-ised instruction. The teachers were free, and apparently felt free, to use as much as or as little of the approach as they chose.

Use of the Fife project materials does require some change in the teacher's role since the booklets provide pupils with much of the information that the teacher might otherwise give. Sampson (1975, p. 122) uses a theatrical metaphor: the teacher moves from centre stage to the positions of the prompter and stage manager. Sampson observes that record keeping becomes important, that teachers may have to learn how to accept situations in which they do not know all the answers, and that teachers must make special efforts to monitor pupils' progress.

The teachers in the SSMCIS pilot schools helped to shape the textbook commentaries, but they played no part in initiating and planning the project or in determining the mathematical content. Teachers' complaints about a given topic usually led to some revision of the treatment but seldom, if ever, to abandoning the topic. Revisions in the original syllabus were debated in the June conferences largely on the basis of personal preferences rather than reports from the experimental teaching. Use of the SSMCIS programme requires the teacher to have considerable mathematical knowledge, but it cannot fairly be said to require a new pedagogical approach. The SSMCIS teacher can teach the new content in much the same way as he taught the old. His classroom role remains the same except that he is teaching more advanced mathematics, and he may be teaching brighter students than he has taught before.

Although Dalin (1973), claims that in Malmö teachers play a large part in the discussion of priorities and in the creation of projects, it appears that, except for Öreberg, they did not play a large role in initiating the IMU project. Teachers were, of course, involved in the field experiments. The results of the experiments, as reported by Larsson, indicated that both pupils and teachers wanted more group teaching (a special component was included in version 5 of IMU to meet this request). Teachers became more involved in individual contact with pupils, as had been intended, but small-group work showed little increase, contrary to intentions. Preparation and planning activities increased, and if the teachers had an assistant, as often happened, they spent less time on routine tasks.

2.3. *The relationship between the centre and periphery*

In the Fife Mathematics Project, curriculum development began at a local centre: the high school in which the pilot experiment was conducted. As the project developed, methods and materials were disseminated to more peripheral schools in the county, which in turn responded with their reactions and suggestions for improvements. From the perspective of the county school system, curriculum development was occurring at the periphery, with financial support and official endorsement provided from the centre. From the perspective of the Scottish educational system, the project was developed at and propagated from the periphery.

The SSMCIS project, from its inception, received financial support from the federal government. Official recognition and grants for teacher training courses helped to encourage local school systems to join the project, but the central government did not actively promote adoption of the SSMCIS curriculum on a wide scale. The local centre at which the SSMCIS curriculum was developed and disseminated was the project headquarters at Teachers College, Columbia University. The pilot schools were not truly the loci of the development work; they were sites for the experimental tryouts only. Dissemination occurred from the local centre to the periphery. Again, from the perspective of the US educational system (or, more precisely, the fifty state systems), the project was developed at and propagated from the periphery. Differences from the Fife Mathematics Project included national financial support and recognition, a project office located outside the schools, and relatively little contribution from the periphery to activities and decisions at the local centre.

The IMU project originated, like the Fife Mathematics Project, with one person working at one school, but like the SSMCIS project, the IMU project conducted most of its work from a headquarters office located in an institution of higher education. The Malmö School of Education served as the project headquarters, the local centre that managed the curriculum development process. The IMU project illustrates how, from the perspective of the Swedish educational system, curriculum development can occur at the periphery and then be picked up for propagation from the centre.

Obviously, what is centre and what is periphery depends upon one's perspective. The very notion of a curriculum development project connotes some kind of centralised organisation: a project is a 'planned undertaking', and planning requires some mechanism for achieving agreement. The very notion of dissemination of curriculum development connotes some movement toward the periphery: that is where the action is. But, as the three illustrative projects suggest, development and dissemination can occur anywhere within a national system of education.

Exercises

1 The method of developing a curriculum at a centre and then disseminating it to the periphery is sometimes advocated on the basis of efficiency. What are some disadvantages of the method?

2 Harlen (1977) distinguishes two dimensions of participation by teachers in curriculum development: (a) a research role, which involves choosing the problems on which to work, and (b) a development role, which involves developing solutions to the problems. She contends that teacher participation in a project can be characterised by either a weak or strong research role and either a weak or strong development role. How would you characterise teacher participation, according to Harlen's distinction, in the three projects discussed in this section? Do you know of a project in which teacher participation can be characterised differently?

3 Dahllöf (1974, p. 164) does not favour teacher-initiated curriculum reform over centralised national reform: 'The different strategies do not seem to lead to the same type of innovation. The teacher initiated reform approach 'from within' is no doubt very effective for bringing about a renewal of the syllabus within a certain subject matter area, but it is doubtful if it can ever lead to any profound change in the total balance of the curriculum, at least not within a limited period of time.' Comment and exemplify.

4 Choose a project known to you and comment upon it using the three headings of Section 2.

5 Re-examine the three projects described above in the context of the perspectives given in chapter 1; for example, consider the different 'stages' in their development.

4

The Practice and Management of Curriculum Development

The case studies in the preceding chapter illustrate something of the variety in recent approaches to curriculum development in school mathematics. The discussion of contrasts between the three projects that was presented at the end of the chapter, however, greatly oversimplified the manifold differences to be seen in curriculum development practice. A more detailed and comprehensive view of such practice requires consideration of the many factors that operate to enhance or constrain curriculum development.

In this chapter we shall examine some factors influencing curriculum development in school mathematics. First, we shall look at curriculum development 'from the outside in', noting forces outside the school and the classroom that shape any efforts to change the curriculum. Then, we shall look at curriculum development 'from the inside out', noting problems faced by the teacher in putting new ideas into practice. Finally, we consider how curriculum development has been managed amid the welter of confusing, and sometimes conflicting, demands for an improved curriculum in school mathematics.

1.   Forces shaping curriculum development efforts

Think of a school mathematics 'classroom' anywhere around the world. It may be an open-sided hut on the plains of Africa, a laboratory room in a primary school in a European city, or a group of pupils working at home in front of their television sets in the Australian outback. The teacher may have a wealth of materials for teaching, only blackboard and chalk, or nothing at all. The class may span many levels of age and ability, or it may be restricted to a narrow range; the course may be required of all pupils, or it may be optional, it may lead to an external examination or it may not. Mathematics classrooms come in a staggering diversity of forms. Nonetheless, they all operate in a society and as part of that society's educational system. Whatever mathematics classroom you thought of can be seen, from the outside in, as the centre of concentric circles representing the local community, the region, and the nation. Within each of these circles, one can identify forces that influence what happens in the classroom.

1.1 *Society's aims and traditions*
The history and culture of a people influence curriculum development in school mathematics by shaping beliefs regarding both the importance of the study of mathematics to that people and the need for change in the mathematics curriculum. We examined some national beliefs about the curriculum in the preceding chapter. How are these beliefs manifested?

Sometimes one can find explicit statements of a country's educational aims.

> To develop in children's minds the Communist morality, ideology and Soviet patriotism; to inspire unshakable love towards the Soviet fatherland, the Communist party and its leaders; to propagate Bolshevik vigilance; to put an emphasis on atheist and international education; to strengthen Bolshevik willpower and character, as well as courage, capacity for resisting adversity and conquering obstacles; to develop self-discipline; and to encourage physical and aesthetic culture.

So the USSR *Great Encyclopedia* of 1948 described the mission of the Russian schools, a mission which, whatever its merits or demerits, would certainly influence the curriculum and the way in which it developed! Since 1948 there have been notable changes of policy in the USSR concerning school organisation and the planning of the curriculum; nevertheless, the basic aims have changed little and the role of the State in fixing and delineating objectives and the methods by which these should be attained has not varied. Education is still directed firmly towards nationalistic ends: social indoctrination and training for participation in the nation's industrial development. The latter aim has, of course, led to considerable emphasis being placed on mathematics and on the selection and education of a mathematical élite (see, for example, Swetz, 1979).

Russia is by no means unique in the stress it places in education on nationalism. Such tendencies can be found in many developing countries where nationalism can help bolster self-respect. In his essay *Education for Self-Reliance,* President Nyerere (1968) argued that since Tanzania would continue to have a predominantly rural economy for a long time, education would have to be devoted to improving village life. The educational system of Tanzania, therefore, should 'foster the social goals of living together, and working together, for the common good . . . [It] must emphasize co-operative endeavour, not individual advancement; it must stress concepts of equality and the responsibility to give service which goes with any special ability, whether it be in carpentry, in animal husbandry, or in academic pursuits.' (p. 52).

Nyerere went on to explain that although the country needed young people who were prepared for the work they would be called on to do in improving their society, 'this does not mean that education in Tanzania should be designed just to produce passive agricultural workers of different levels of skill who simply carry out plans or directions received from above. It must produce good farmers; it has also to prepare people for their responsibilities as free workers and citizens in a free and democratic society.' (pp. 51–52). Developed and developing countries alike may stress education's instrumental value in promoting the attainment of nationalistic aims.

Education is also seen in many countries as a means for achieving and cementing national unity: as it was in France and Germany. To some extent this was also the case in the USA, where there has always been a need to assimilate large numbers of immigrants. Yet education in England and the USA has traditionally differed from that in France and Germany. In England more emphasis has been placed on the personal development of the individual; in the USA education has often been viewed as a preparation for citizenship, and less emphasis has been laid on book-learning. Again, education in Germany has been strongly influenced by the redoubtable academic tradition of its universities, whereas in France the need to produce a cultured, educated élite has often appeared paramount.

Traditional views of education (even when expressed in such black and white terms as above) cannot be lightly cast aside or changed. The teachers of today are the (successful) products of yesterday's system and are imbued with its characteristics.

The emphasis on logic, orderliness of mind, and clear thinking associated with the French – *ce qui n'est pas clair n'est pas français* – characterises their 'modern maths' texts in the same way in which a pragmatic approach does that of their English counterparts. When

Burke *(Reflections on the French Revolution)* remarked that he could find 'not one reference whatsoever . . . to anything moral or anything politic; nothing that relates to the concerns, the actions, the passions, the interests of men,' he could well have been reviewing a French modern mathematics text.

Views on education are clearly represented in a nation's educational structure. Thus it is not surprising that the USA and the Scandinavian countries should have adopted comprehensive education, whereas in France significant changes to the multi-track system have been resisted.

Some countries have a history of educational innovation and of progressive educational thought: Switzerland, the USA, England, and Germany. In others, the opportunities and encouragement have not always been present, or it has been the custom to look elsewhere for ideas. Thus Australia has often looked to Britain and the USA for guidance. Progress can best be made in a climate in which educational innovation is expected and encouraged. That this might in fact lead, without adequate safeguards, to over-ambitious and ill-advised changes is no excuse for inertia.

Unfortunately, many developing countries have to fight to free themselves from the aims and traditions that were built into their educational systems by their colonial masters. Often the reaction has been to impose a nationalistic element on to a system which in the words of Nyerere was 'developed regardless of our particular problems and needs'. Not every nation has realised with him that there is an urgent need *'to think . . . about the education we want to provide'.*

Some of the strongest forces for curriculum reform in developed countries over the past two decades have emerged from the movement toward comprehensive secondary education. The movement has compelled a re-examination of educational aims.

> Mathematics and science, closely interlocked, are the basis of the most revolutionary of recent developments in society and in the everyday lives of all young people. Even the slowest pupils are interested in progress and success, and in demonstrating that mathematics can contribute towards success, we may best hope to give all pupils before leaving school some realization of its intrinsic value.

This paragraph is from the Newsom Report (1963, p. 151), which dealt with the education between the ages of 13 and 16 years of pupils of average or below average ability in England. One can see in the paragraph quoted not only the concern for 'a change of heart' regarding the pupils who are characterised in the report's title as 'half our future', but also an expression of belief in 'progress and success' as important stimuli for the study of mathematics. Although the Newsom Report does not give an explicit rationale for teaching mathematics to secondary-school pupils of ordinary ability, it clearly suggests that mathematics will contribute to 'education for all' if its usefulness can be shown and if pupils can find success in using it.

The authors of the Newsom Report expressed the desire to bring all pupils 'to an interest in the content of mathematics itself at however modest a level' partly through the successful use of mathematics as a tool. Giving all pupils success in mathematics, however, conflicts to some extent with one of the traditional functions that mathematics has performed, that of sieve or 'gatekeeper'. In many countries only those pupils who are 'successful' in school mathematics are allowed entry into higher education and into the professions. Educational authorities require the study of mathematics for a variety of reasons having to do with traditions, beliefs about liberal education, beliefs about the utility of mathematics, and so on, but in practical terms school mathematics has historically fulfilled an important role in helping authorities identify those qualified for society's rewards. Although a country may be willing to extend the benefits of secondary education to a larger fraction of its population, it cannot change the criteria for 'success' in school mathematics without either increasing the opportunities for higher education and entry into the professions or abandoning school mathematics as gatekeeper. The perception of mathematics as a difficult and demanding discipline, by pupils, parents, and teachers alike, has been important in maintaining school mathematics in its gatekeeping function. Curriculum developers who want to make the study of mathematics more attractive and more easily learned should be aware that their efforts may run counter to traditional conceptions of mathematics that serve practical, if unacknowledged, purposes (see, e.g. Revuz, 1978).

Exercises

1 Find statements of national objectives for education in your country (British readers may have trouble with this question!) and contrast them with the statement from the USSR *Great Encyclopedia* quoted earlier and with Nyerere's statement (1968, p. 53) that education must 'encourage the development in each citizen of three things: an enquiring mind; an ability to learn from what others do and reject or adapt it to his own needs; and a basic confidence in his own position as

a free and equal member of the society, who values others and is valued by them for what he does and not for what he obtains'. What are the implications of each statement for the school mathematics curriculum?

2   'A 'standardised test' is an implicit statement of objectives.' (Howson, 1979). Discuss. On what bases and by whom are 'standardised tests' constructed?

3   Krulik & Weise (1975) suggest the following as goals for mathematical education.
(i) To achieve for each individual the mathematical competence appropriate for him.
(ii) To prepare each individual for adult life, recognising that some students will require more mathematical instruction than others.
(iii) To foster an appreciation of the fundamental usefulness of mathematics in our society, particularly with reference to understanding and improving man's environment.
(iv) To develop proficiency in using mathematical models to solve problems.
How do these subject goals support/conflict with the national objectives mentioned in Question 1 above?
What contribution can/do such lists of goals make to curriculum design?

4   According to Nyerere (1968, pp. 46–7), education in Tanzania under the colonial government 'was not designed to prepare young people for the service of their own country; instead it was motivated by a desire to inculcate the values of the colonial society and to train individuals for the service of the colonial state. . . . The educational system introduced into Tanzania by the colonialists was modeled on the British system, but with even heavier emphasis on subservient attitudes and on white-collar skills. Inevitably too, it was based on the assumptions of a colonialist and capitalist society. It emphasized and encouraged the individualistic instincts of mankind, instead of his co-operative instincts'.
What practices in school mathematics teaching might lead to the development of 'subservient attitudes'? Can you find evidence of such practices today?
What practices in school mathematics teaching might emphasise and encourage pupils' 'individualistic instincts'? Can you find evidence of such practices?
Outline a seven-year primary school mathematics curriculum that would prepare pupils for life and service in the villages and rural areas of a country such as Tanzania. If you can, compare your outline with syllabuses actually used in such a country.

1.2.   *Social and geographical factors*

An educational system is governed to a considerable extent by social and geographical factors and these too will greatly affect curriculum development and its management. For example, Norway and Sweden are remarkable for their unified value and social structures and are relatively small, predominantly rural, and prosperous. The USA is marked by conflicting social and ethnic interests, values, and views on national priorities; is large; and contains wide extremes of prosperity and environment.

As a result, in the Scandinavian countries it is possible to envisage curriculum development as an integral part of a national policy of socially-oriented educational reform. In the USA, this is not possible. Although small enough to be treated as an entity, the Scandinavian countries are also sufficiently large and prosperous to be able to manage curriculum development internally. Moreover, although their languages do not have international currency, as English and French do, they are sufficiently 'strong', and educationally and culturally well-established, to remove any doubts as to the desirability of basing school instruction upon them.

Conditions are very different, therefore, from those which exist in, say, Botswana and Swaziland. These two African countries have each populations roughly equal to that of Oslo, and under half that of Stockholm. (Yet, Botswana is larger in area than any European country with the exception of the USSR.) The two African countries are poor, they lack a tradition of schooling, and such schools as they possess are largely geared to overseas standards and objectives. Moreover, the size of their school population is such as to make the preparation and provision of specially designed secondary materials extremely expensive: publishers can never have large sales and it would seem almost impossible to offer teachers a choice of suitable texts.

Curriculum designers, then, face far greater obstacles than do their colleagues in developed countries. In addition, syllabuses must be geared not only to a five-year course leading to the Cambridge Overseas School Certificate, but they must also dovetail with a local Junior Certificate examination, taken after three years of secondary education with the dual role of terminal examination and sieve. Again, and we shall return to this in a later section, a country's examination structure can be a major factor in curriculum development.

Another difficulty faced by educators in most African states is that of language. Instruction, particularly in the secondary sector, is often given in a foreign language, English or French, rather than in the child's mother tongue. Indeed, the latter will often differ from the national language, the medium, possibly, of primary education.

Language problems also arise in many developed countries. The USSR is vast and has no common vernacular: children are instructed in their mother tongue until they have proceeded to a point where Russian can be introduced as a second language. In, for example, Belgium and Canada there are socially divisive alternative languages of instruction; in Britain and the USA there are communities of immigrants whose children are linguistically handicapped at school.

In some countries, there are religious or ethnic divisions within the educational system. All such factors will affect the way in which curriculum development can take place.

1.3.   *The professional status and training of teachers*
The professional quality and the degree of autonomy granted to teachers varies considerably from country to country, and somewhat surprisingly the two are not always linked. Thus, for example, the teachers in Scottish schools have traditionally been better qualified, both academically and professionally, than their English counterparts. Like their legal and medical colleagues they have a professional Council (which English teachers do not) which regulates standards and, *inter alia,* guards against any professional misconduct. Yet in the classroom they enjoy less freedom than do English teachers, for the Scottish system is more centralised, the teacher is offered fewer choices, and the inspectorate adopts a patriarchal rather than an avuncular air.

As has been mentioned, in the past the education of primary school teachers everywhere has often been academically impoverished – and that of secondary schoolteachers professionally non-existent. Now the need for both academic and professional training is being recognised, and teaching is on its way in some countries to becoming an all-graduate profession.

Yet whilst professional training is accepted as vital, there are differences in how that phrase is interpreted. In England, say, stress is laid on the practical aspects of teacher training, on periods of practice teaching in schools. Education departments in Germany, on the other hand, appear to place more emphasis on history, philosophy, and indeed, theoretical considerations in general. This difference matters not only at the initial training level, but it also colours the teacher's view of university educators and pedagogues: are they interested in the everyday problems of schools and have they anything to contribute towards their solution? This question assumes great significance in the processes of curriculum development.

The quality of the teachers concerned is probably the key factor in any classroom innovation. This, however, depends on factors other than pre-service training, important though that is. There must also be provision for in-service training, for the constant up-dating of professional knowledge. Incentives for teachers to keep abreast with developments must be provided. Inevitably, this will lead to short-term, financially-based solutions, such as efficiency bars which deny academic and financial advancement to non-participants (as in Czechoslovakia), or even limited accreditation periods (as in some of the United States). More important and effective, however, would be the creation of a climate in which such in-service education comes to be expected as the norm. The provision of financial incentives and efficiency bars may well prove to be means to that end, but this greater objective must not be lost sight of, and it must be borne in mind that it is a mark of a professional that he is prepared (in both senses*!) to make decisions and to accept responsibility for them.

1.4.   *The educational system*

The degree of 'state' as opposed to 'local' control of education varies greatly from country to country. France and Sweden are examples of highly centralised systems in which the curriculum is firmly controlled from the centre. Thus in Sweden the study plan and syllabus, including the way in which the weekly hours are to be divided, is centrally determined: the teachers however, being given latitude for manoeuvre.

Centralisation can also take place at a state rather than a national level as is the case in Australia and West Germany. In some cases, e.g. the USSR and GDR, there is national control, but administration is divided. The USA is marked by the emphasis on local control. As in Australia, the federal government has no responsibility for, or direct control over, education, although it does make a financial contribution. In England control is divided between the central Department of Education and Science and the eighty or so 'local education authorities'. These bodies do not, however, determine the curriculum of the school, which is, subject to certain constraints, in the hands of the headteacher and his staff.

* Prepared: *brought into a suitable state, trained* and also *disposed, willing.*

Control over the curriculum can, therefore, be exercised at several levels, and the degrees of freedom granted to the teacher can vary substantially. In particular, we note that the system for determining the curriculum may be different from the system for administering and financing the schools.

Yet it must be stressed that the actual influences on the curriculum are always more numerous than the official system would suggest. In countries like Sweden and France, where the curriculum is centrally controlled by means of syllabuses and handbooks of methods for teachers, there is always a gap between official prescriptions and actual practice, a gap which becomes especially pronounced when new syllabuses are first mandated. No matter how 'centralised' the system, an individual teacher, once the classroom door is closed, assumes some degree of autonomy. Conversely, in countries like England and the USA, where the curriculum is nominally a local matter, social and political pressures act to impose conformity. In other words, centralised systems are not so centralised and decentralised systems are not so decentralised, as commonly supposed. As a French school inspector once observed: 'In France, every teacher is supposed to be doing the same thing at the same time but nobody is, and in England, where everyone is supposed to be going his own way, nobody is'. (Koerner, 1968, p. 52.)

Exercises

1   The traditional local control of education in the USA may be undergoing some change. Mosher (1977) contends that 'in assessing the current state of American federalism the question of the locus of effective power – who dominates in the partnerships or cooperative arrangements – is a matter of considerable dispute' (p. 119). What are some factors that have contributed to a greater federal and state role in US education? Can similar shifts of control be found in other countries?

2   The Open University Curriculum Design and Development Course Team (1976) identified three groups of European countries having similar administration organisations for education and consequently similar approaches to curriculum development: (a) the Scandinavian countries; (b) France, Belgium, Austria, and Spain; and (c) West Germany and the Netherlands. They characterise the first group as having little decentralisation but some central government movement towards it, the second group as having little decentralisation but some grassroots movement towards it, and the third group as having

considerable decentralisation but some myth-making about local autonomy.

How valid is this grouping scheme?

Where do the following countries fit into the scheme: Italy, England, USSR, USA?

1.5.   *External examinations*

A major constraint in the teacher's autonomy – not necessarily for ill – is the external (public) examination system. The Prussian *Abitur* (p. 21) and the French '*bac*' were in existence by the start of the nineteenth century. Their influence has grown rather than waned in the intervening period, and the latter in particular exerts a near stranglehold on the country's education. Public examinations began in England in the 1850s. For just over a century their influence was restricted to the 'academic' child. Since 1964, however, public examinations have been made available to the great majority of school pupils.

A feature of these newer (CSE) examinations has been the opportunity they have provided for schools to draw up their own syllabuses and for teachers to be responsible, with some outside moderation, for designing and operating the assessment procedures. This feature has, to a limited extent, been available to German teachers for many years.

Significant questions that must be asked of any country's public examination system are as follows.

(i) Do schools have a choice of syllabus? Can they opt for, say, a 'modern' syllabus as opposed to a 'traditional' one?

(ii) Who controls and administers the examination? Is this done by a State department as in France or by private bodies as in England?

(iii) How easy is it for teachers to bring about syllabus changes? Indeed, what part do schoolteachers (as opposed to inspectors, university teachers, etc.) play in the examining process, for example, designing syllabuses and setting and marking examination papers?

(iv) What modes of assessment are used? Are there opportunities for continuous assessment, for projects or investigatory work? Are examinations written or oral or a mixture?

(v) *Quis custodiet ipsos custodes?* Who examines the examiners?

Exercises

1   According to Nyerere (1968, p. 62), the preparation of young people for the realities and needs of Tanzania 'requires that examinations should be down-graded in Government and public esteem. We have to

recognize that although they have certain advantages – for example, in reducing the dangers of nepotism and tribalism in a selection process – they also have severe disadvantages too. As a general rule they assess a person's ability to learn facts and present them on demand within a time period. They do not always succeed in assessing a power to reason, and they certainly do not assess character or willingness to serve.'

What are some ways of overcoming the disadvantages of external examinations that Nyerere identifies?

2   In many countries external examinations are the principal means for universities and other institutions of higher education to influence the secondary school curriculum. Has the extent of this influence changed in recent years? How have academics influenced the secondary school mathematics curriculum of pupils who will not go on to higher education?

3   Describe the examination system of your country. (Include answers to questions (i) – (v) above.)

1.6.   *The inspectorate*

The inspectorate would seem to be a feature of nearly every educational system. Originally, the inspector was a somewhat rough and ready evaluator of the teacher's competence and the pupils' progress. In Victorian England the money allotted to a school depended on the results achieved by its pupils in a test administered by one of the roving band of Her Majesty's Inspectors. Since that time the inspectorate's function has changed considerably! In England the HMI has the dual role of monitor of the educational system, in which he provides a service to the Ministry, and of adviser and encourager of teachers. The task of 'reporting' on individual schools and teachers has virtually disappeared. The many local education authorities will also employ inspectors or advisers, but here a conflict of interest can arise. A local adviser may not only have to advise but also to assist in the appointment of teachers to posts of responsibility carrying higher salaries. Advice when given under such circumstances has a somewhat different connotation!

In some countries the role of the inspector is less equivocal. He instructs and formulates rather than advises, and wields considerable power in matters of promotion and within the examination system.

Clearly, within the field of curriculum development it matters greatly whether the inspectorate are impartial onlookers (as in England), leaders of a partnership with teachers (as in Scotland), or enforcers of a self-designed curriculum.

In the United States, which does not have a formal inspectorate, the educational system is monitored in various ways. Local and state boards of education monitor the use of funds, the performance of the staff, and the quality of a school's graduates. The quality of a secondary school's curriculum is assessed most directly by accrediting agencies, which visit a school every few years to assess the qualifications of its staff, the nature of its facilities, and its instructional programme. Accreditation is necessary if a school wishes to have its graduates accepted into colleges and universities, but in practice accreditation once received is seldom lost. The quality of instruction is monitored indirectly by state departments of education, who certify teachers to teach, and directly by district supervisors, school principals, and department heads, who both advise and encourage teachers and, especially during the probationary years, observe their teaching. The increasing use of tests in the USA to ensure that teachers and pupils meet 'minimum standards of competence' can be seen as an attempt to compensate for the absence of an inspectorate.

1.7.   Texts and publishing

The textbook continues to be a major influence on the classroom: in many cases it still effectively determines the curriculum. How texts are written and selected for classroom use is, therefore, of paramount importance in curriculum development.

The situation varies from country to country. At one extreme are countries with a single centrally written and approved text which all schools must use (as in East Germany), at the other those, such as England, in which the teacher selects freely from a wide range of commercially-produced texts. The former system provides the teacher with no choice, yet with a text that is usually professionally and mathematically competent to a high degree even if it is of an overall bureaucratic dullness. In the latter, mathematical and pedagogical soundness may well yield pride of place to gimmickry both in content and design. The commercial temptation to supply that which teachers will find easy to use will be strong indeed.

In between these two extremes come a variety of other possibilities: a choice from two state-approved and state-produced texts (Poland), or from a number of state- (or board-) approved, commercially-produced texts (USA and Canada); that of using *free* copies of state-approved, commercially-produced texts or of *buying* texts yet to be approved (Federal Germany).

The delicate problem of state involvement in school publishing has arisen on several occasions in connection with innovation. Thus when the Schools Council for the Curriculum and Examinations was established in England it was on the basis that it would not publish pupils' materials. The fear of state publishing was temporarily allowed to outweigh the fact that the production and publication of such materials is an almost essential feature of successful innovation. In Sweden the government bought a share in a publishing house, an action which resulted in the argument being made that 'the link between a State publishing house and a highly centralized system would become an obstacle to fruitful innovation if it literally gave an imprimatur to new orthodoxies and restricted the choice of materials available to schools'. (Maclure, 1971.)

In the USA a survey (Weiss, 1978) of the use of materials from projects supported by the National Science Foundation found that, whereas many of the textbooks produced by science curriculum development projects were being used in schools in 1977, relatively few textbooks produced by mathematics curriculum development projects were still in use, although they had been used in previous years. Various explanations for this difference can be offered, but one of the most compelling arises from the observation that more science-project textbooks than mathematics-project textbooks were brought out in commercial editions.

Commercial textbook publishers in the USA have set up elaborate networks for the promotion and distribution of their products. Teachers in the USA depend heavily on textbooks. This dependence has roots that go back to the introduction of free public education into a frontier society: the textbook was used to compensate for a shortage of well-educated teachers. As a result, textbooks are marketed much like automobiles: publishers' representatives find out what the consumer wants and then the publishers compete to offer it to him in the most attractive package possible. Novelty is at a premium; yet the changes offered are frequently cosmetic and rarely radical. The greatest changes in US textbooks in the last three decades have come from outside the commercial arena – many of them in response to the stimulus provided by federally-funded projects – but in each case the forces of the marketplace acted quickly to dampen the change.

Exercise

In what ways did the following innovative textbooks influence, or fail
to influence, the mathematics curriculum.
a. *Éléments de Géométrie* (1741) by A. C. Clairaut.
b. *Basic Geometry* (1940) by G. D. Birkhoff and R. Beatley.
c. *Mathematics: A New Approach* (1962) by D. E. Mansfield and
   D. Thompson?

The lesson to be drawn from the observations in the preceding section
of this chapter is that curriculum development activities take place in a
context largely determined by society. Curriculum developers cannot
ignore the social, political and educational systems in which the school
curriculum is embedded, nor can they expect to transplant successfully
curricula that have worked well in some other context. Countries that
could not afford to develop their own materials have often attempted to
import ready-made curricula from other countries (see chapter 8). Such
imports seldom survive the move, not simply because the countries and
their educational systems are different but also because their views of
the curriculum are different. Various approaches to managing curriculum development have arisen in response to particulars of time and
place. How well they might work in another context depends, in part on
the similarity between the two contexts: that of the donor and that of the
recipient.

2. The teacher's role in curriculum change

The teacher is the mediator between the curriculum and the
child, and any attempt to change the curriculum must consider the
teacher's role. Teachers can be implicated in curriculum development in
two related ways: as participants in the process or as users of the
product.

We saw in chapter 3 that teachers in the Fife Mathematics Project
collaborated with the project organiser in setting the direction for the
project and that eventually he stepped out of the picture. Teachers
played an important role in trying out the workcards, but they did not
do as much creating of materials as in projects such as the School
Mathematics Project and the Mathematics for the Majority Continuation Project. The teacher's role in curriculum development varies from
project to project within a country, but it varies even more between
countries, primarily because of national differences in expectations for
teachers and in conceptions of their responsibilities.

In the USA, curriculum development tends to be product-oriented, and the teacher is cast in the role of consumer. Teachers may participate in the process of curriculum development, but they are usually treated more like guinea-pigs than partners. Curriculum developers may recognize the need to get teachers committed to a new curriculum if it is to be successful, but the developers see the problem more as 'selling' the product rather than getting involvement in the process. The notion of a 'teacher-proof' curriculum has found currency in the USA but virtually nowhere else.

In England, curriculum development tends to be more process-oriented, and the teacher more often takes the role of creator. The English hold to the myth that the teacher is responsible for determining the curriculum. As Maclure (1972, p. 41) has noted, this is a myth

> in the sense that it expresses great truths in a form which corresponds more to an idea than to reality. The less factually correct it may be, the more important it is to assert. . . . To refer to this as a myth is not to denigrate it. It is a crucial element in the English educational idea. It is the key to the combination of pedagogic, political and administrative initiatives which provide the drive for curriculum reform in England and Wales.

The role of the teacher in official curriculum development projects is shaped by beliefs as to who can and should determine the curriculum. In Sweden and Spain, politicians and administrators are seen as responsible for the curriculum, and teachers play a correspondingly minor role in the curriculum development process. In the USA, teachers are seen as appropriate participants in the process, but they are often considered incapable of making a significant contribution. In England, participation by teachers is not only appropriate but expected.

The impact of curriculum development on the teacher's role is also a function of beliefs as to what is possible and appropriate. Developers who believe teachers should not be given responsibilities that properly belong to state educational authorities tend to produce curriculum materials that offer the teacher few options although they may demand a high level of teaching competence. The new Soviet mathematics curriculum for secondary schools appears to be of this nature (Maslova, Kuznetsova, & Leont'eva, 1977). Developers who believe teachers incapable of contributing to the curriculum development process tend to produce materials that bypass the teacher, reducing his role to that of monitor or clerk. An example is the Individually Prescribed Instruction – Mathematics Program developed at the University of Pittsburgh (Ovsiew, 1973). Innovators who believe teachers are the key curriculum developers tend to elevate process above product; the materials these developers produce are often no more than samples or source materials that require the teacher to do his own developing. Examples of such materials are the teachers' guides produced by the Nuffield Mathematics Project and the Mathematics for the Majority Project, and the teaching units produced by the Unified Science and Mathematics for Elementary Schools Project (see the case studies in chapter 6).

Consideration of the practice and management of curriculum development requires attention as to how these processes affect the teacher's role. In chapter 3, for example, we noted that the Fife Mathematics Project moved participating teachers from centre stage to the wings and the prompter's box. Such shifts in role may be difficult for teachers to make. Sociologists such as Lortie (1975) have observed that most teachers get their greatest rewards as professionals from the actual instruction of pupils. A teacher feels most rewarded when he has 'reached' a pupil or group of pupils and they have learned. When teachers move off the instructional stage, they need to find new sources of satisfaction. Some teachers have been quick to adopt individualised systems such as the Fife Mathematics Project and the IMU project because they believe that by shifting much of the instructional burden to workcards and workbooks, they will have more time to spend helping pupils who are having special problems in learning. When management of such programmes in the classroom begins to interfere with opportunities to work with pupils, however, teachers are likely to revert to more traditional modes of instruction (*cf.*, for example, Larsson, 1973).

The school culture in any country is highly resistant to change and especially to change introduced from outside the school. When 'outsiders' are able to institute curriculum changes without the cooperation and collaboration of teachers, such changes seldom go smoothly or last long. Sarason reports the case of a school system in the USA in which the mathematics supervisor and members of the board of education attempted to introduce a 'New-Math' curriculum programme in grades 4 to 6 of the district. Teachers were not involved in making the decision to change curricula. After the decision had been made, they were informed of the impending change and were invited to attend summer workshops to learn about the new programme. Sarason (1971, p. 44) reports on what happened during the school year:

> Teachers generally were anxious, angry, and frustrated; many children were confused and many parents began to raise questions about what

was going on and about their own inability to be of help to their children; parent workshops were organized. It was an unsettling year for practically everyone involved.

Small wonder, then, that in subsequent observations of elementary school classrooms, Sarason and his colleagues found little evidence that the goals of the curriculum reformers had been met.

It is important to note that the problem of effecting curriculum change is not solved simply by making sure that teachers are involved in the process from the outset. In a fascinating series of case studies, Stake and his colleagues (1978) report their findings after interviewing teachers of science and mathematics (Stake *et al.*, p 16: 1).

> What we learned from many of our direct interactions with teachers in this study was that they were not just taking a 'sour grapes' attitude about curriculum improvement. They were not cool toward innovation just because they were not the ones invited early to participate in curriculum development programs or institutes. They had been telling anyone who would listen that they *know* what will work in their classrooms, and what will not, and that they know that most of the heralded innovations will only work in exceptional situations.

The teachers in these case studies had strong beliefs that one could not separate subject-matter aims of instruction from aims relating to the socialisation of pupils. They thought of pupils as needing certain ways of being handled before they would get to work and certain kinds of materials before they would learn. Attempts to change the subject matter, its organisation, or its mode of presentation have to be accommodated to teachers' beliefs as to what good teaching is and what it is reasonable to expect of pupils.

From inside the classroom the teacher sees the mathematics curriculum as only part of the educational process. Although many of his goals involve imparting mathematical knowledge and developing mathematical abilities – and he judges his success in large measure by how well his pupils have learned the mathematics he has tried to teach them – he has other goals that relate to schooling in general. He will change his teaching only if he sees the change as warranted by these broader goals as well as the narrower goal of teaching mathematics.

Even if the teacher wants to change the curriculum and his teaching, in line with his goals for mathematics teaching in particular and schooling in general, he may not be able to sustain a change if other features of the school, such as administrative structures, time for planning, and money for equipment and supplies, are not changed.

Many projects in which teachers have engaged in curriculum development activities have been successful as long as the project could supply assistance and lend moral support to participating teachers. But when assistance and support have been withdrawn, teachers have returned to their previous activities. This retrogression is especially true of projects, such as Unified Science and Mathematics for Elementary Schools, that have made relatively heavy demands on teachers' time and energy.

An important lesson that every successful teacher must learn is that to the pupils who are learning it, the subject matter does not look the same as it does to the teacher. The pupils are in the midst of learning, whereas the teacher has presumably learned more and can look down from a higher vantage point and with a different perspective. It helps a teacher to attempt to assume the pupil's view from time to time. An important lesson for curriculum developers is that to the teachers who are teaching it the subject matter does not look the same as it does to the curriculum developer. The view from inside the classroom is different, and to be effective, curriculum developers must come inside.

Exercises

1 Researchers who study teachers and teaching often comment that teachers are, and feel, very much alone in their work. As a profession, teaching does not seem to support collaborative arrangements. What are some ways in which the mutual isolation of teachers can be reduced? Can you cite examples of curriculum development in mathematics that has come from collaborative work by teachers? How has this work been sustained?

2 Stake and his colleagues (1978) found that mathematics seems to have *ritual* value for parents and teachers: it helps the school impart values such as persistence, neatness, thoroughness, and diligence. What are the implications for curriculum change of the mathematics teacher's role as custodian of this ritual value?

3.   Managing curriculum development

The history of the mathematics curriculum as an object of attention by educators has been sketched in chapter 2. We have seen how, before the rapid expansion in educational opportunities, the school mathematics curriculum changed slowly and usually only in response to the innovative ideas of individuals who managed to win acceptance for new topics or approaches. Even then many changes were often ephemeral. Sometimes, the incorporation of new ideas into textbooks helped to make the changes more lasting. For example, the textbooks of Warren Colburn (1821) helped translate Pestalozzi's ideas to America. Curriculum development, therefore, usually occurred as the expression of one person's desire to make a planned change in instruction. The need for 'management' did not arise.

A dawning awareness of the school mathematics curriculum as something to be studied and purposefully developed, rather than simply being allowed to evolve, can be seen in the activities of the early subject associations and the CIEM (chapter 2). The establishment of the latter in 1908 prompted many countries to look critically at their curricula, and since that time numerous special committees and commissions have been organised to study various curriculum problems. In some countries the groups were set up or supported by the Ministry of Education, and these frequently led to revised curricula, for example, in Prussia in 1922 and Ontario in 1924. Elsewhere the initiative lay with subject associations and examining bodies.

Most of the early groups concerned with the revision of mathematics curricula limited their work to producing reports. These often contained detailed recommendations concerning curricula, but were rarely accompanied by teaching materials. Neither was it customary to try out the recommendations in any systematic way. An important development occurred in 1951 when the colleges of education, engineering and liberal arts at the University of Illinois set up a joint committee (UICSM, see chapter 6) which attempted not only to develop a curriculum plan, but also to produce text materials based on classroom tryouts and to train teachers in their use. UICSM, then, offered what came to be recognised as characteristic features of a *project*: trials, published materials and in-service education. It represented one solution to the problem of how the curriculum development process could be *managed*, of how innovation could be effected. The model used was one which, as reference to the Lockard Reports (p. 272) shows, was quickly adopted throughout the world.

The advent of the project served to focus attention on the need for the management of curriculum development, something which could not be avoided when people began to call themselves curriculum developers and when curriculum change began to be seen as a continuing (and complicated) process rather than an occasional event.

Moreover, as the shortcomings of the original project model gradually became apparent, various alternative organisational frameworks were established for the coordination and advancement of developmental activity and for the dissemination of innovation. We shall now examine some of these forms in more detail.

Exercise

The 1922 German curriculum was a revision of work begun in 1905 by Felix Klein and his collaborators. Investigate Klein's curriculum and its revision. Can you find parallels with recent curriculum proposals?

As we shall see, the early 'management' model to be used by projects was one imported, almost unchanged, from industry. However, its shortcomings soon became apparent as the full significance of 'centre–periphery' considerations and complications were revealed. Gradually, various attempts were made to provide, through the establishment of agencies, networks of centres, laboratories, institutes, etc. organisational frameworks for the coordination, advancement and dissemination of innovation. The variety of forms of management that has resulted in the different countries is, indeed, bewildering: further evidence, if that be still needed, of the complexity of the problems involved and of the absence of obvious solutions.

3.1. *The 'centre' and the periphery*

Regardless of how the responsibility for the curriculum is allocated within an educational system, such a system tends to be organised into a network with a centre from which channels radiate out to schools and classrooms. This centre may belong to a national ministry of education, or that of a state or region.

The illustrative projects in chapter 3 showed how the locus of curriculum development can vary. The curriculum can be determined and/or developed at the centre and then transmitted through the channels to the periphery, as is done when a government not only supports the development of a curriculum but mandates its use. The curriculum can also be developed at the periphery and then transmitted to the centre, as occurs when teacher-made materials are put into a central collection or bank. Another possibility is that the curriculum is developed by a project outside the system and then spreads through the system either from the centre or from points further out towards the periphery.

The 'centre → periphery' model was the one favoured by most of the early projects: a small group of competent specialists combined to develop new course material which was then made generally available to teachers. It is the way in which, for example UICSM, SMSG, SMP, Nuffield Mathematics, and the Mathematics for the Majority Project operated. These examples illustrate that even within this one model there can be vast differences in approach. The two last-named projects, for example, provided teachers' and not pupils' materials; the SMP was run by practising schoolteachers and was not financed from public, i.e. state, sources.

In contrast, the Mathematics for the Majority Continuation Project (MMCP) placed more emphasis on the 'periphery'. It was centrally administered and the materials were centrally planned and edited, yet the detailed writing and experimentation was done by a large number of working groups throughout England, Wales, and Northern Ireland. Thus writing took place away from the centre and involved many teachers. A somewhat similar plan was followed by the Caribbean Mathematics Project (CMP). It, however, attached even more weight to 'peripheral' activities.

Further consideration of the examples we have given will reveal that attempts to classify projects according to whether they are 'centre → periphery' or 'periphery → centre' are unlikely to prove very fruitful. In practice, the arrows are nearly always two-way. Yet we are presented with a spectrum of possibilities, and a rough ordering of projects according to the weight given to central and peripheral activities can be made: for example, UICSM, Continuing Mathematics Project (UK), SMP, MMCP, CMP.

Centralised educational systems typically propagate a curriculum from the centre, but they may not develop it there, relying instead on selected experimental institutes and schools on the periphery. Decentralised educational systems may develop curricula at the centre, as examples that local authorities might want to imitate or that individual schools might choose to adopt; they may develop curricula outside the system; or they may depend entirely on the periphery for curriculum development. In some instances, for example, in West Germany, the delineation of syllabuses and the development of curricula can be divorced; the former made a 'central' activity, the latter left to those on the periphery.

A curriculum developer may have no choice as to whether the educational system he works in is centralised or decentralised, but he often can control the points where the curriculum is developed and whence it is propagated through the system.

Exercises

1 Schaffarzick (1975, p. 247) makes the following argument against local curriculum development. 'Teachers and most district personnel have neither the time, the energy, nor the expertise required for sustained work in developing or continuously revamping entire curricula. Furthermore, work spread across hundreds of schools and districts would lead to much unnecessary duplication. And it is virtually impossible for school districts to pull together even a semblance of the resources (scholars and other experts, as well as finances) that are possible with a pooling of funds.'
   How serious are these objections? How might they be met?

2 Choose a curriculum development project with which you are familiar and locate the point or points in the educational system at which the project was developed and from which it was propagated (if it was propagated). What would have been some likely effects on the project if other points in the system had been used instead?

3 Draw diagrams to illustrate the form of 'centre – periphery' model adopted by particular projects. Indicate in each case where initiation, administration, writing, testing (developmental work), revising and editing took place, and show the routes of 'disseminatory flow'.

3.2. *Forms of management*

A variety of structures has been devised for managing the curriculum development process, and these structures do not lend themselves easily to classification. Three broad categories do, however, suggest themselves. Many curriculum development activities are conducted in and through what are usually called *centres*. (We refer here to centres for curriculum development, not centres of educational systems, as in the previous section.) A centre is characterised by a certain amount of autonomy in operation. Although development activities may take place in numerous schools affiliated with the centre, the activities are managed by the centre staff. Sometimes the curriculum development process is divided among centres that are organised into a *network,* a second category of structure. Some networks have one centre that coordinates the work of the others; other networks consist of coequal centres. But in each case, the centres have distinct objectives and perform somewhat different functions. The management of either a centre or a network, on the other hand, might be external to the structure. Management structures of this sort can be termed *agencies.* An agency manages curriculum development activities that occur elsewhere. Let us consider some examples of each kind of structure.

a. Centres. Most curriculum development projects of the last two decades have been conducted at centres set up especially for the project. Such centres are usually located at a college or university. For example, the School Mathematics Study Group (SMSG) was managed from a centre located first at Yale University and later at Stanford University (Wooton, 1965). The School Mathematics Project (SMP) began as a research programme of Southampton University (Thwaites, 1972), later moving to Westfield College of the University of London, and then becoming an independent charitable trust. The Caribbean Mathematics Project (CMP) (Wilson, 1978) had its base at the School of Education of the University of the West Indies. The 'alef' Elementary Mathematics Project was located at the University of Frankfurt.

These 'centres' were attached to existing institutions. In other cases centres have been geographically independent and some have catered for a wider range of curricular activities. For example, the Education Development Center, Newton, Massachusetts, has managed not only USMES and the Arithmetic Project Course for Teachers, but also projects in science and social science. The Curriculum Development Centre of Sri Lanka, which has conducted projects across the curriculum, including ones in elementary and junior secondary mathematics, is but one example of a national centre.

An important national centre concentrating solely on mathematics was established in the Netherlands: IOWO, the Institute for Development of Mathematics Education (IOWO, 1976). This has undertaken projects in elementary and secondary school mathematics (Wiskobas and Wiskivon, respectively). IOWO sought to produce curriculum 'documents and procedures' with maximum participation by teachers and others who were to be involved in the dissemination process; in-service education being an important part of its approach.

IOWO's curriculum development strategy was formulated in an effort to avoid mistakes made in earlier projects. In fact, however, there are many similarities between IOWO's strategy and those used previously. A major difference is in IOWO's extensive involvement of teachers and other educators outside the project in helping to shape the proposals for the curriculum. In this respect, as in others, IOWO stands between projects such as SMSG, SMP, and USMES and another type of curriculum development centre: the teachers' centre.

This type of centre arose as a result of the work of the Nuffield Mathematics Project which established such 'teachers' centres' in the pilot areas in which its programme was being tested.

The centres were intended to provide not only workshop courses for teachers but also the opportunity for teachers to meet, discuss their work, and collaborate in developing teaching ideas and materials. 'These centres have grown and multiplied, and in so doing have cast off their original Nuffield labels and become general "curriculum development centres" concerned with all aspects of the curriculum and with both primary and secondary education. They are supported financially by the local education authority' (Griffiths & Howson, 1974, p. 149).

The concept of the teachers' centre has spread to many countries and has broadened to include a variety of activities. Teachers' centres range from school rooms set aside for teachers' work after school to separate buildings given over to classrooms, laboratories, and collections of materials. Their activities range from a modest programme of one or two meetings a term to intensive collaborative work by a group of teachers in developing the curriculum for a school course. Teachers' centres usually have someone who serves part-time (or even full-time) as director, secretary, or warden, organising and managing the centre's activities. Some type of centre organisation is clearly the most common structure for managing curriculum development work, with the exception of the activities of individual teachers and textbook authors. Centres seem to provide the minimum amount of structure needed for people to work together on a curriculum development project.

b. Networks. In some countries attempts have been made to establish a network of 'centres' each undertaking specific activities as part of a national or regional effort: the advantage being that research and developmental work is distributed widely, thus ensuring that no teacher feels too distant from a curriculum centre, whilst still being coordinated so as to avoid wasteful duplication. In France, for example, considerable work has been done at the Instituts de Recherche pour Enseignement des Mathématiques (IREMs). There are about thirty IREMs, one at almost every university. They were begun in 1969 to provide in-service training for teachers to help them implement a new national mathematics syllabus for secondary schools. The IREMs have not only offered courses but have also undertaken various research and development activities (see, for example, Revuz, 1978). The Lyons IREM, for example, has written a modern secondary school course that combines textbook and worksheet material (see Bell, 1975).

In the USA, 21 research and development centres and 20 regional educational laboratories were established between 1965 and 1967 in the wake of the Elementary and Secondary Education Act of 1965. A decade later only nine centres and eight laboratories had survived the heavy weather of diminished federal spending and increased criticism regarding federal support of educational research and development activities.

The research and development centres were intended to bring together scholars to develop systematic programmes of research and development dealing with a common area of concern, such as higher education, teaching, or evaluation. Each centre was based at a university and was supposed to bring together resources and interdisciplinary talent to attack a significant educational problem. The centres usually began their work by supporting a set of separate projects, each somehow related to the common problem. The successful ones tended to be those that developed some system for facilitating teaching and learning, a model or prototype that could be developed and field-tested by the centres or by cooperating regional laboratories. Examples of such systems include Individually Prescribed Instruction (IPI), a product of the Learning Research and Development Center at the University of Pittsburgh, and Individually Guided Education (IGE), a product of the Wisconsin Research and Development Center for Individualized Schooling at the University of Wisconsin.

The regional educational laboratories are primarily engaged in development activities rather than research. They were set up to provide a link between the research and development centres and the public (state) schools. Each laboratory is a non-profit corporation with its own board of directors, staff, and other sources of income. The laboratory is supposed to serve the educational needs of its region (regional boundaries were never defined precisely, 'region' merely meant 'more than one state'), but no successful laboratory has had the resources to handle more than six or so projects at a time. Laboratories have concentrated on developing curriculum materials and programmes for teacher training that are based on work done by the centres and other researchers. For example, Research for Better Schools, Incorporated, the regional laboratory in Philadelphia, developed, expanded, and disseminated the IPI programme begun by the Pittsburgh centre.

Federal support and supervision of the laboratories and centres are now carried out by the National Institute of Education, which was established in 1972. The role of the National Institute of Education in promoting curriculum development is still under debate, however, and the 'network' of laboratories and centres does not have the organisation and coherence the term implies. At present, the laboratories and centres form a loose kind of association that conducts some innovative curriculum development work, with special emphasis on 'basic skills'.

c. Agencies. As curriculum development projects multiplied and the need to keep the whole of the curriculum under review was recognised (a need which subject-centred projects are incapable of meeting) so gradually a third kind of organisation, the agency, came into being. Agencies are distinguished by the fact that, in general, they support and fund curricular activities rather than organise them.

In England and Wales, the major agency is the Schools Council for Curriculum and Examinations, or Schools Council, which was established in 1964 (see chapter 6). The Schools Council is an autonomous body composed of representatives of teachers' unions, local government associations, higher education, the Department of Education and Science, employers, parents, etc, which is funded by both central and local government.

In addition to its projects, the Council has sought reforms in examination systems and techniques and has encouraged the establishment of teachers' centres. Unique features of the Schools Council's approach are the relative freedom from governmental control, the extent of teacher participation in development, and the variety of activities undertaken, which concern virtually all aspects of the school curriculum and range from modest trials of new techniques to large-scale projects to develop course materials. The amount of power exercised by the teachers' unions, in particular that to which primary and secondary modern schoolteachers traditionally belonged, was the cause of considerable criticism and led in 1978 to a reconstitution of the council.

The major supporters of educational research and development in the USA are the National Institute of Education, the Office of Education, and the National Science Foundation, in that order, at least according to total dollars spent. The National Science Foundation's Division of Pre-College Education in Science supported most of the curriculum development projects in school mathematics in the USA during the 1960s. In recent years, as the number and extent of curriculum development projects has diminished, the National Institute of Education has been the main supporter of school mathematics curriculum development projects, primarily through the regional educational laboratories and the research and development centres.

These government agencies have tended to use a scheme of grants and contracts in which funds are given to an institution – usually a college or university, but sometimes a non-profit-making organisation such as a professional society – to support the development or dissemination of materials to improve the curriculum or the instructional process. The early projects funded by the National Science Foundation tended to be directed at revamping entire courses and arose primarily from unsolicited proposals to the Foundation. Recently, with funds more scarce and the initial wave of reform past, government agencies such as the Foundation have tended to 'target' their funding, calling for competitive proposals on one or more themes, such as the integration of science and mathematics, or mathematics for the student who will not go into a scientific field. Projects have been supported to produce units, or modules, that can be fitted into existing courses, as well as to produce materials for complete courses. Relatively more funds in recent years have been given for projects to disseminate existing materials to teachers.

Competitive proposals to an agency are evaluated through a system of peer review, in which a panel of reviewers from the disciplines concerned reads and rates the proposals that have been submitted. Proposals are presumably funded according to merit, but when merit is equal, geographical distribution is sometimes taken into account. Once materials have been produced, and an account given of how funds were used, either the materials are entered into the public domain or commercial publishers are given an opportunity to bid for the right to publish them.

In Sweden, the National Board of Education has been the sponsoring agency for curriculum reform projects funded by the government. The Board was reorganised in 1964, and since then has played the main role in coordinating educational research and development in Sweden. As Vormeland (1973, p. 429) observes, 'Few school systems and school reforms have been founded to the same degree on educational research as have the Swedish systems and reforms'. The Board has been able to disseminate research and development activity relatively quickly into schools through its curriculum guides and teaching aids. Special agreements with the government-owned Educational Publishing Company help reduce the lag between development work and production. At the beginning of the 1960s most of the research activity sponsored by the Board was initiated and conducted by researchers in institutes at colleges and universities, but the Board has increasingly taken more initiative in organising and conducting research projects, and the scale of these projects has increased. The IMU project was begun at the request of the National Board of Education.

Many other countries have similar agencies, within the Department or Ministry of Education, that are responsible for supporting curriculum development and that may be charged with both continuing the process of curriculum innovation and sponsoring research that might feed the innovation process. Australia has the Curriculum Development Centre, Scotland has the Consultative Committee on the Curriculum, Norway has the National Council for Innovation in Education, and so on. These agencies vary in the amount of funding and autonomy they enjoy, and the power they have to implement their programmes varies according to the degree of centralisation of the country's educational system. But it appears that once these agencies have been established in response to some perceived national need, they tend to take an increasingly stronger hand in guiding and coordinating curriculum development activities. It seems to be natural for such agencies to seek new territory and to begin to anticipate, rather than simply respond to, problems arising in the schools.

Exercises

1 'Both the English and Scottish systems can be criticized for their relative weakness in making an impact on schools and teachers. Diffusion and implementation fall short of expectation, possibly because the Schools Council projects can too readily become professionalized development work which seems alien to the classroom teacher, while the CCC reports are not easily distinguished from the hortatory pamphlets to which the teacher has developed a protective immunity' (Nisbet, 1976, p. 169).

   Comment on Nisbet's assessment of the Schools Council's and the Consultative Committee on the Curriculum's activities. In what ways, if at all, does mathematics prove an exception? Nisbet warns that both systems rely too much on the 'expert', rather than pursuing problems as seen by the practitioner. Do you agree?

2 The peer-review system for judging the merit of proposals for curriculum development projects has been attacked on the grounds that the stronger institutions get a disproportionately large share of the awards and that the government has some obligation to assist weaker institutions in the interest of future national strength. The system has been supported on the grounds that science must rely on the verdicts of expert peers. If you were setting up a granting agency, would you give grants for curriculum development on merit alone even if they

were not evenly distributed across institutions? What role would you give to peer-review?

3   We have mentioned the dramatic decline in the fortunes (and numbers) of the US research and development centres and the regional educational laboratories. At the time of writing IOWO's existence as an independent institution is about to end and the IREMs are likely to be shorn of all but their research role. Investigate whether similar economies have been made in your country. What reasons have been given for these cuts in the support for curriculum development?

4   Ruddock & Kelly (1976, p. 98) suggest that 'the distinction between profuse and confined systems is . . . more useful when considering dissemination than that between centralized and decentralized systems'. Here, a 'profuse' system is one which contains a variety of development and dissemination agencies (e.g. Netherlands, Federal Germany, England); a 'confined' system 'is characterized by the limited variety of agencies they contain, the intimate connections that exist between development and dissemination and the lack of necessity for co-ordinating agencies because, in a sense, the systems are self-co-ordinated' (e.g. France, Denmark, Ireland).

How would you classify your educational system? What differences concerning dissemination would one expect to find in 'profuse' and 'confined' systems?

5   Consider the extent to which the choice of different forms of management, centres, networks, agencies, has depended upon geographical and political considerations.

6   In many countries 'curriculum development' is administratively divorced from 'teacher education' (including in-service education). What is the situation in your country? What difficulties arise from such a division of duties and how might they best be overcome?

7   The Schools Council has responsibility for curriculum and examinations, but not for teacher education or 'fundamental' (i.e. not specifically curricular) research. The IREMs had responsibility for research and in-service education, but not for determining curricula and examinations.

Investigate the 'responsibilities' of other agencies, networks etc., and comment on the effects of the ways in which these are circumscribed.

8   Investigate changes over the last two decades in the policy of the National Science Foundation regarding procedures for commercial publication of NSF-supported curriculum materials. Does government support of curriculum development inhibit commercial publishing activities? Does it put publishers at an unfair disadvantage?

3.3.   *Project management*

The project is the major contribution of the second half of the twentieth century to the field of curriculum development. It arose as technological society's response to the problem of making a qualitative change in the school curriculum, and it was patterned after engineering projects that had brought teams of scientists and engineers together in crash programmes to develop weapons, break codes, etc. The early curriculum development projects borrowed their strategies, consciously or not, from the procedures used by industries in developing new products. This is undoubtedly a major reason why such projects began originally in countries having a high degree of local autonomy in curriculum making and only later appeared in countries with more centralized systems.

Projects such as UICSM, SMSG, and Nuffield mathematics have been characterised as following the R–D–D (research–development–diffusion or dissemination) model (see, for example, IOWO, 1976). Actually, however, few projects used a true R–D–D approach. Their development activities were seldom founded on a research base (either created or borrowed), and their diffusion activities were often rudimentary unless and until the diffusion process was taken over by commercial publishers.

A key to the organisation and life of a project is the director. Most projects have been born from the initiative of one person, who became the director, and have been organised to accommodate the director's abilities and interests. Few projects of the nineteen-fifties and -sixties survived a change in directors, and few directors found the energy to head more than one curriculum development project. Some directors, such as Begle of SMSG, Fehr of SSMCIS and Thwaites of SMP, did very little of the actual curriculum development themselves. Their talents lay in recruiting and inspiring others to do the initial development and then in organising tryouts, revisions, and evaluations. Other directors, such as Beberman of UICSM, supplied many of the pedagogical ideas themselves but relied on others for ideas on mathematical content. Still other curriculum developers, such as Papy and Dienes, provided both pedagogy and mathematics but did not organise and direct teams of other workers. In general, the larger and more comprehensive the project, the more likely it was to have a director who gave 'managerial' rather than 'conceptual' leadership.

The director has often been responsible for setting up the project in the first place, selling the idea to an agency that could provide sponsorship and support. In some cases, such as SMSG, the idea for a project arose from a meeting or conference, and a director was found to get the project under way. In other cases, as, for example, the three projects discussed in chapter 3, it was the director who supplied ideas for meeting an apparent need. Getting and keeping financial support for a project is a non-trivial part of the director's role, and occasionally this and public relations work have occupied most of the director's time. In some countries, the director's principal task may be to win official recognition for the project. He may have to demonstrate to the satisfaction of the educational authorities in the country that the project will yield an improved mathematics curriculum. He may be asked to provide evidence that pupils and teachers will respond well to the innovation, or it may be that the testimony of experts will suffice, providing they are the right sort of experts from the authorities' point of view.

The larger projects have also tended to have some sort of advisory board, consultative committee or council which works with the director in formulating policy and setting the direction for future work. The advisory group has usually been set up to satisfy the funding agency, governmental or private foundation, that its money was being spent in a responsible way. Some advisory groups have had only a symbolic function, but most have made substantive contributions to the direction of the project, and a few have managed to counter some of the director's initiatives.

Many of the projects characterised as R–D–D projects operated in a manner like that of the SSMCIS (see chapter 3): teams of mathematicians and mathematics teachers wrote material which was tried out in pilot schools, revised and tried out again one or more times, and then distributed to a larger audience. In a variation on this theme the writing team was composed entirely of mathematics teachers. In most cases the teachers in the pilot schools either were on the writing team themselves or met with the writers to suggest revisions.

Pilot schools were seldom selected at random; they were usually invited to join the project because mathematics teachers at the school were known personally or by reputation to members of the project. Although this method of setting up a network of pilot schools ordinarily ensured cooperation and easy communication between the school and the project staff, it also meant that the pilot schools were likely not to be typical of the schools to which the final project materials would be disseminated. In projects having a large number of pilot schools, much of the director's and staff's energy had to be given to the 'care and feeding' of the pilot schools.

As noted earlier, projects in the USA have tended to treat the teacher as a consumer of the curriculum rather than its creator. To an extent this attitude is built into the R–D–D approach, but it can be seen clearly in other projects as well, including some outside the USA. Whenever a project makes use of expertise and resources that are not available to typical schools, it will tend to offer teachers a 'prepackaged' curriculum. Projects differ considerably in the demands they make on teachers in terms of both mathematical and pedagogical sophistication. The SSMCIS project materials, for example, required a knowledge of mathematics that was seldom possessed by junior and senior high school mathematics teachers in the USA. The implementation of the SSMCIS programme, therefore, demanded some means for teaching the mathematics to the teachers, and special in-service courses were offered, not only to the teachers in the pilot schools, but also to teachers in school systems that were planning to adopt the scheme. Needless to say, any project that requires extensive in-service education of this sort will encounter problems in diffusing its programme and materials. In the case of projects such as Nuffield Mathematics and USMES, the demands on the teacher were more pedagogical than mathematical. Nuffield wanted teachers to employ a 'discovery' method of teaching and hoped to provide them with the necessary help through the teachers' centres. USMES wanted teachers to stimulate the investigation of 'real' problems; it used workshops to train teachers to serve as leaders in their school systems. The amount and type of training which a project offers teachers depends upon both its philosophy regarding teacher participation in curriculum development and its view of the special training its programme requires.

In a sense, every project has to 'sell' itself. Even if the project is only a group of local teachers getting together to share ideas, the participants must be assured of the value of the activity for them. If the project is a major national effort to revamp the teaching of mathematics at some grade level, the selling job is more difficult since more people have to be convinced that the new is better than the old, that the proposed innovation is worth the trouble of changing. People tend to judge a project's success by its ability to win acceptance for its ideas. As we noted before, however, 'new' does not necessarily mean 'better', and teachers who decide not to accept a proposed change may be quite justified in their decision. A broader criterion for the success of a project might be its ability to stimulate teachers to reflect on their work.

Every teacher, then, is a consumer of curriculum development work. In the USA, this role is obvious, and the marketing of a curriculum development project's wares requires attention not only to their pedagogical merits but also to the competition they will encounter in the marketplace. In other countries, the role of the teacher as a consumer may not be so obvious, because he is expected to do more of the creative work himself and/or because he has fewer alternatives to choose from, but the role is there. When a state mandates a new syllabus, a new examination, and an accompanying textbook, the teacher is nonetheless in the position of a buyer. If he does not buy the philosophy behind the new syllabus, if he does not agree with the changes in the examination, if he does not see the point of the new content in the textbook, he will end up teaching a curriculum at some remove from the official one. The management of any curriculum development project, whether or not it plans formal dissemination activities, should seriously consider how the project's ideas will be implemented in classrooms.

Exercises

1 Develop criteria for evaluating the 'success' of a project.
2 Select a project of which you approve and present a case intended to convince an educational authority of its value.
3 Investigate the phenomenon of the 'pilot school'. How is it, or how does it become, different from the 'average' school?

4.   Determinants of curriculum development practice and management

This chapter has stressed the social context of curriculum development and how this context determines what form the innovation will take and how it will be managed. The social context is the critical 'outside' force shaping curriculum development. But there is an 'inside' force as well. That force is the content of the innovation, the ideas it attempts to embody and the philosophy that motivates it. Ultimately, the critical inside determinant of curriculum development practice and management in school mathematics is the view innovators have of mathematics itself and the reasons for teaching it in school. When mathematics has been seen as a body of knowledge to be acquired for practical reasons, curriculum development practice and management have aimed at the efficient acquisition of this knowledge. When mathematics has been seen as important and interesting in its own right, developments have depended heavily upon the contributions of experts in mathematics who shared this view and curriculum development has been managed so as to acquaint teachers with it. Comprehensive, organised programmes have usually been the goal, if not the result. When mathematics has been seen as valuable for its applications to other fields, curriculum development practice has been organised around the creation of curriculum 'units' that can be 'plugged in' to existing schemes. The creation of completely new programmes built on applications has been much less common. When curriculum developers have been concerned more with the process of doing mathematics than with the knowledge of mathematics, they have attempted to engage teachers in the parallel processes of doing mathematics themselves and developing a mathematics curriculum for their pupils.

Curriculum development in mathematics, then, has both a context and a content. Both determine its structure, its operation, and its achievements. It is to a more detailed consideration of the content of curricular reforms that we now turn.

5

Curriculum Theory and Curriculum Research

1.   **The legacy of early curriculum theories**

The early decades of this century saw a rapid expansion in the provision of secondary education in the United States. Public high school enrolments, which stood at about 500 000 in 1900, roughly doubled in each of the following four decades.

There was a need then to reconsider school curricula which had to that time not differed significantly from European educational tradition. America needed its own, newly-developed pattern.

The result was an autonomous reform movement in pedagogy, based on a pragmatic philosophy, which was to determine the basic reorientation of American curricula. In the field of mathematics education two opposing tendencies emerged: one originated in social practice; the other viewed social practice as the goal of its reformatory intentions.

The most significant educational figure to emerge was John Dewey (1859–1952). Dewey derived his conception of learning from his observations of how everyone learns in his environment, i.e. through action and experience. Learning in schools should, therefore, also be conceived of as 'learning by doing' and 'learning by experience'. Dewey rejected the division of content into separate school subjects, since this was alien both to the child and to reality. Instead he proposed project teaching based on real objects. He saw teaching founded on children's activity and experience as being in keeping with reality in two vital respects: it takes into account the natural requirements and needs of the child's development and also provides the best guarantee that the child will become fitted for life outside school.

Dewey's writings heralded the 'progressive era' of American pedagogy in the 1930s and 1940s which, however, reduced his ideas to the process-oriented cultivation of 'good learning activities' in schools. 'Curriculum' was interpreted so widely and, in a technical sense, extensively that it became synonymous with life and experience.

The second tendency, that of restructuring the curriculum to adapt it to the requirements of practice, became linked with the 'behaviourist' approach.

Here the starting point was Taylor's job analysis which sought to provide a theoretical basis for the introduction of assembly-line work and job evaluation. Taylor's ideas were generalised by Bobbitt in *The Curriculum* (1918) and *How to Make a Curriculum* (1924), the resulting thesis being that all activities necessary in society should be classified according to a certain number of job characteristics, to which abilities and skills correspond. These have then to be considered as the goals of school instruction.

Bobbitt's theory was extended and exemplified in detail by E. L. Thorndike in his *Psychology of Arithmetic* (1922). In this, Thorndike developed the instruments of behaviourist learning theory by applying the principle of 'taylorisation' of more complex activities within social practice to the learning process, school arithmetic being the area within which he chose to substantiate his theory. In particular, by giving a structure to the learning process, which he identified with that of social practice, especially within industry and commerce, a functional connection between employment and education was established. This connection was, however, not without its accompanying dangers. Orientation to the general needs of social practice could be distorted to yield an emphasis on a narrow range of specific techniques identified by employers.

This work, based on behaviourist psychology, marks the first of a series of approaches to the problems of curriculum development, approaches which we shall identify and later use as guidelines when discussing the reforms of 1950–75. The work of Dewey and Bobbitt, however, also had implications for the direction in which further progress in general curriculum theory was sought in the United States.

Bobbitt's interest had, of course, not been simply that of making a curriculum, but of answering the question of how to make a curriculum, i.e. of finding a general solution to the problems of curriculum development. When, shortly after World War II, the discussion on curriculum theory was resumed, the new contributions inherited this orientation: an intermediate position between practice and theory. As a consequence, when curriculum projects began to be established, they were unable to base their activities on such research: it was too abstract. (It must, however, be stated that those involved in the early projects showed little awareness of the existence of a body of theory.)

Bobbitt's practical interest was process engineering. That means, he saw the problem of curriculum development as one of responding to an unspecified variety of needs and goals; a problem which should be solvable given a means of systematic classification. This was an idea which was to have a long-standing appeal to theorists, fascinated by the presumed benefit of such a systematic organisation of the basic field of curriculum development.

In 1949, Ralph Tyler referred to earlier work, but he rejected the thesis that either individual *or* social needs should determine the curriculum. He claimed that equal consideration should be given to individual, social and cultural needs (the latter to include the demands of science). This triad of needs was to be generally adopted by the American school of curriculum theory.

Tyler, himself, placed considerable emphasis on a scientific orientation. He proposed a new categorisation of curricular goals, successors to the famous 'Seven Cardinal Principles' of education postulated by the National Educational Association in 1918. The arrangement (see p. 88) resembled that of Bobbitt's, but was, of course, better fitted to mid-century life.

This conception of Tyler's was followed by many authors, who sought to refine and elaborate the proposed system.

One such author was Hilda Taba, an educator greatly influenced by Dewey. She contributed further suggestions on the role of science in the curriculum, pointing out that both content and methods of science must find a place in school education. Her main interest, however, was in demonstrating how the child's needs should determine curricular decisions. As subjects for analysis she included the nature of knowledge and its formation in the cognitive process, socialisation and educational influences outside school. Thus she came to postulate that careful decisions on goals, content, organisation and evaluation of a curriculum could only be made on the bases of comprehensive theories of society and cognition, and on a comprehensive anthropology.

Yet Hilda Taba accepted that theoretical speculation would not by itself suffice, and attempted to overcome the abstractness of theoretical reflection by designing curricular modules.

The developing categorical system of curriculum development was further enriched by J. I. Goodlad, who investigated the manner in which decisions relating to curriculum development are taken. He correlated different types of decisions, concerning values, educational aims, educational objectives, learning opportunities and conditions of learning, with different authorities, social, institutional and individual.

By this time, however, curriculum development was proceeding apace and, in order to connect theory to current practices, Goodlad proposed empirical studies on how curriculum development actually occurs. He thus initiated a new trend in curriculum theory, namely, to abandon (at least partly) a theory claiming to be prescriptive, in favour of one with a descriptive, explanatory function. This view of curriculum theory was to become more important as growing experience of curriculum development resulted in important procedural changes.

This move towards empirical stocktaking, coupled with Taba's request for a more thorough underpinning, reflected contemporary unease about the development of curriculum theory. An enormous structure had been erected, yet its increased 'perfection' did not make it better fitted to practice. However, there is no indication that this failure was felt to be a result of shortcomings within the theoretical construct itself. Was a categorical framing on the basis of the need-response scheme really appropriate to the problem?

Exercises

1 Try to find old schoolbooks of the progressive era and/or accounts of the 'project plan'. Investigate how the ideas of Dewey were translated into the mathematics curriculum.
2 'Education is a process of living and not a preparation for future living'.
   Discuss this statement of Dewey's and its implications for schools.
3 Which modern projects owe most to Dewey? Are there significant differences in their and Dewey's conceptions of education? (Dworkin (1959) and Garforth (1966) summarise Dewey's writings.)
4 The following classifications for the construction of curricula have been offered:
   (a) by the 'Committee of Ten' of the National Educational Association, 1918: (I) Health, (II) Command of fundamental processes, (III) Worthy home membership, (IV) Vocational training, (V) Citizenship, (VI) Worthy use of leisure time, (VII) Ethical character. (Neagly & Evans, 1967.)
   (b) by Bobbitt: (I) Language activities, (II) Health activities, (III) Citizenship activities, (IV) General social activities, (V) Sparetime activities, (VI) Keeping oneself mentally fit, (VII) Religious activities,

(VIII) Parental activities, (IX) Unspecialised or non-vocational activities, (X) The labor of one's calling. (Eisner, 1967)
(c) by the Virginia State Curriculum Study: (I) Protection and consideration of life, (II) Natural resources, (III) Production of goods and services and distribution of the returns of production, (IV) Consumption of goods and services, (V) Communication and transportation of goods and people, (VI) Recreation, (VII) Expression of aesthetic impulses, (VIII) Expression of religious impulses, (IX) Education, (X) Extension of freedom, (XI) Integration of the individual, (XII) Exploration. (Tyler, 1949)
Consider the bases on which these principles were established. Compare them, and discuss the place which mathematics would hold in the three conceptions of the curriculum.

2. Style and theory

At a conference at Allerton Park in Monticello, Illinois, in September, 1971, educators from ten countries met to discuss styles of curriculum development (Maclure, 1972). One participant, R. A. Becher, asserted that recent approaches to curriculum development could be divided into three styles: instrumental, interactive, and individualist. The instrumental style is exemplified by the UICSM project and by other projects initiated in the United States in the late 1950s. Projects of this style were begun in order to increase the number and preparation of students pursuing careers in science and mathematics. Becher saw projects in the interactive style developing in the mid-1960s, when the curriculum reform movement spread to other school subjects and began to aim at other populations of students. The emphasis of these projects was on social interaction and cooperation. By the end of the 1960s, projects in the individualist style had begun. Such projects emphasised self-development, with the teacher cast in the role of assisting the student in learning. Becher argued that the three styles migrated one after another from the United States to Britain and Scandinavia and then to Western Europe, with each nation progressing through the sequence at its own rate. Becher's analysis is too neat to stand without argument, and it apparently received considerable discussion at Allerton Park. A separate proposal by Georges Belbenoit noted three perceived needs that curriculum development can attempt to meet: economic efficiency, social justice and democracy, and individual and collective satisfaction. These needs match Becher's styles fairly well, as do the three educational goals identified by one of the discussion groups at the conference: productivity, equity, and self-realisation. These correspondences lend some support to Becher's analysis, but the conference participants were not entirely satisfied with it. Becher used a matrix to characterise his model, and by the end of the conference he had modified the matrix substantially in the light of the discussions. The modified matrix given by Becher & Maclure (1978) is shown in Table 1. Any curriculum development project can be matched against the matrix by choosing the description in each row that best fits the project. The Allerton Park conference participants attempted the task with Becher's original matrix and were successful in fitting some projects into a column, but found other projects spread across columns. As Maclure, author of the conference report, noted: 'The temptation is to use the matrix as the first stage of a parlour game. There is no limit to the number of rows or columns which could be added. Nor is there any possible way of constructing verbal boxes which every project can be made to fit neatly.'

The difficulty of fitting projects to proposed clusters was but one factor that led conference participants eventually to conclude that it is impossible to identify and characterise such styles so that they can be

Table 1. *Curriculum development styles* (Becher & Maclure, 1978)

	Cluster I Instrumental	Cluster II Interactive	Cluster III Individualistic
View of knowledge	Packages (subject disciplines)	Problems (interdisciplinary inquiry)	Personal exploration (eclectic searches)
Categories of goal emphasised	Job/career	Social adjustment	Personal happiness
Means adopted	Highly structured materials	Loosely structured (but researched) materials	Unstructured (non-existent?) materials
Teachers' classroom roles	Dominating	Managing	Assisting
Dissemination strategies	Teachers as passive (rational) recipients	Teachers as representative (token?) participants	Teachers as (partial?) developers
Evaluation techniques	Attainment of pre-specified goals	Anthropological ('illuminative') studies	Individual case-histories
View of humanity	People as things (manipulable)	People as social animals	People as individuals
View of external reality	*Terra firma* (the real world)	Sandbanks (the changing world)	*Terra incognita* (the unknowable, therefore unknown)
	Newton?	Einstein?	Berkeley?

understood and followed. Instead, they tended to favour lists of questions that can be asked about a curriculum development project, with the answers used to categorise the project along a variety of dimensions. Such an approach does not presuppose the existence of styles, but it does allow patterns to be seen. A paper prepared for the conference by the Schools Council in London illustrates the approach. The following questions were asked of each of 16 Council projects:

> Why have the projects been set up?
> What sort of aim do projects have?
> What are the outputs of projects?
> How do projects set about their work?
> What sort of people undertake project work?

The answers were used to identify five broad reasons why projects have been established, three main aims, four outputs, and three ways projects have worked, but no stylistic patterns emerged from this analysis.

What happened at Allerton Park? Although it might be an exaggeration to accuse the participants of indulging in 'parlour games', an astonishing fact remains: interest seems to have been solely concentrated on establishing a classificatory system, with little attempt being made to explain existing projects or to understand their theoretical background. The problem was reduced to that of 'fitting projects to clusters'.

This raises the question of the reasons why theoretical research in the field of curriculum development is, or should be, undertaken. Or, expressed somewhat differently, what should distinguish theoretical research from general abstract speculation? A simple answer may well suffice here: theoretical research goes back behind phenomena and provides instruments for understanding. In curriculum development terms, it explains causes, reasons and implications of phenomena, it gives criteria for evaluation and legitimates decisions. The paramount objective of curriculum development theory, however, is to recognise the complexity of inter-dependent factors and to organise them in a manner which will facilitate the attainment of intended goals.

An identification of 'styles' helps us to recognise that specific, characteristic features of curricula may normally occur in clusters. That allows one, say, to infer (with some caution) unknown features of a project from a known one; e.g. if the primary role of the teacher in the classroom is to *assist* in the learning process, then, according to Becher, the category of goals most emphasised may concern the 'personal happiness' of the student. It may have contributed to the frustration of the Allerton Park Conference that the epistemological effect of such a realisation is so limited.

Of course the scope of Becher's structure must not be compared to the theories of Tyler and some of his followers, nor must his intention. We nevertheless chose this example because it illustrates some typical, shared features. In particular, there is the attempt at a systematic approach, and a confidence in the classification of phenomena. We have already pointed out that this may be seen as a tradition going back to Bobbitt. In any case, the consequences of this inclination in American curriculum theory are considerable.

The foundation of Bobbitt's system had been evidence. But the more the determinants of the curriculum were studied, the less could his position be maintained. The development of theories of cognition and the new claims of mathematical science necessitated revised conceptions, starting from the very base of available knowledge and incorporating modern contributions. Such a design was, in fact, developed by Bruner, but not within the school of general curriculum theorists.

Compared with the results of theoretical research carried out in close connection with curricular developments, the categorical approach seems to have been more of a handicap than a help. It dissects phenomena and studies their parts individually, whereas the major problems in curriculum development arise from the inter-relationship of these. Moreover, a systematic classification, once established, tends in practice to be treated as an *a-priori* framework. Fundamental problems, which cannot find a place within the system, will not easily be resolved.

Let us illustrate this with an example. We recognise three categories of needs – individual, cultural and social – and consider the problem of how to introduce fractions in the classroom. In particular, we ask how the time available should be shared between (I) the initiation and development of cognitive processes, and (II) training in computational skill. This question itself raises many others: How does this affect the individual both in terms of motivation and future benefit? What are the interests of society? Who identifies these and using what criteria? What are the wider educational consequences of our decision? Clearly, we shall have to study the requirements and forecasts of cognitive theories, and we shall have to examine how societal demands were identified in a manner which has allowed them to be represented in concrete, instructional form. Finally, we shall have to recognise that all the interpretations, ours as well as the theorists' we study, are very closely related to philosophical and political views on the individual and society.

Obviously, the answers to such problems lie outside the reach of categorical systems of curriculum theory, the justifications for which tend to become less obvious as their specifications grow more elaborate. The result of such self-confinement in theoretical perspectives has been the identification of classification with theory.

It was in an attempt to escape from such constraints that Hilda Taba was prompted to ask for a new approach to curriculum theory. Others, too, have sought to resolve this dilemma, and to regain *terra firma* within actual practice. In West Germany, S. B. Robinsohn set out to provide information on what the 'demands of social and individual life' really are. He started empirical investigations into particular concrete situations defined by applications of mathematics within industry and management. The intent was to analyse what general, non-professional mathematical aptitudes were required if the individual's ability to act autonomously and competently in such situations was accepted as an educational goal. This led to a new view of how one might interpret the legitimation of curricular goals.

In Britain the work of Bernstein bore certain similarities to that of Goodlad. Bernstein's interest, however, was not primarily to develop principles for the construction of curricula. Rather, he conceived his studies in curriculum theory as but one aspect of socio-cultural change, so opening up curriculum theory to a sociological approach, and indicating at the same time the interdisciplinary nature of curriculum theory. Bernstein traces framings and classifications from the observation of existing curricula in order to obtain criteria for comparison, differentiation and evaluation.

These, then, represent attempts to collect data in fields which have been the subject of much discussion, but in which we are now aware that we know very little. They indicate a need for more, and better founded, arguments which would support, but of course not replace, basic theoretical understanding and interpretation.

Exercises

1 Select a mathematics project with which you are familiar and match it against the matrix in Table 1. Which rows of the matrix cause the most trouble?

2 Investigate Becher's assertion that Britain, Scandinavia and Western

Europe progressed at their own rate through the sequence of his three styles.

3   In the preface to the Allerton Park conference report, Maclure asserts 'The basic problem in curriculum development today . . . is to reconcile efficiency and humanity. In the last analysis the curriculum determines what goes on between the teacher and the child and thereby transmits to the child the values of society . . . The very notion that the curriculum must be "developed" in a systematic manner lends itself to technocracy.' Discuss, paying particular attention to the last sentence.

3.   Approaches to curriculum development

We have observed how, from about 1950 onwards, the demands upon, and position of, general curriculum theory in the USA gradually changed. Numerous projects were initiated, many of which were either based on their own theoretical research or adopted research work from related disciplines. As a result, a number of different theoretical tendencies emerged, partly simultaneously, partly in reaction to earlier ones. The study of these tendencies is useful for a variety of reasons.

First, so much work took place in the period from 1950 to about 1975 that there is a risk that the reader will be totally confused. The recognition of different stages of development will, we hope, facilitate orientation. In fact, phases of curriculum development are discernible, each characterised by the dominance of certain theoretical trends manifested in typical projects. This is not to deny, however, that the boundaries between the various approaches are very fluid and the chronology far from exact. The approaches differ more in the emphasis they place on certain determinants of curriculum development than in the selection and diversity of these determinants. Accordingly, projects may well exhibit a blend of approaches or may display dissimilar tendencies at different stages of their work.

Secondly, the different theoretical approaches to curriculum development in this reform period may be compared to the different characters in a play. When we try in retrospect to draw conclusions, it will be useful to remember the contributions these characters made and the themes they accentuated. Unfortunately, we shall not be able to discern continuous progress: the last project does not represent the acme of achievement. Neither is it very likely that any other project will meet all our criticisms. Future action, then, will have to be planned bearing in mind all the various aspects illuminated by the theoretical approaches we describe.

In this chapter we shall give a concise account of these theoretical conceptions, and in the following one will offer a retrospective view of the reform period and of some of its most important projects. As should be obvious from our earlier discussion on categorical classifications, our identification of trends is intended to serve heuristic purposes only: it is not our intention to establish a new systematic structure. The introductory theoretical description does, however, mean that we avoid the temptation of failing to give due consideration of theory in favour of empirical descriptions.

The identification of theoretical approaches was conceived, and in chapter 6 is initially exemplified, in terms of curricular activity within the USA. Similar trends are, however, usually observable in European and other countries, although perhaps delayed, partially obscured, or on a smaller scale, for the US developments had great international impact in the 1960s. More recently the flow of ideas has been less restricted and, in particular, English and Scandinavian contributions to international discussion have found more acceptance. Nevertheless, a study of developments in the USA can do much to clarify approaches, trends and attitudes elsewhere.

3.1   *The behaviourist approach*

The behaviourist approach is aimed at improving learning, that is, at achieving a more effective mathematical education, through a reform of the methods of learning and teaching. The underlying theory proceeds on the assumption that any learning process can be described in terms of a (bond of) stimulus–response pattern, that the learning process can be initiated by constructing a suitable stimulus–response programme, and that the outcomes of learning processes can be 'objectivised' as observable changes in behaviour. In this programme, the learning objective (goal) determines the desired change of behaviour, which can be checked. The success of the learning process is, therefore, open to objective control, for the attainment of the objectives, expressed in terms of behavioural changes, can be readily verified: simple objectives are attained through individual and separate stimuli, more complex ones through the addition of simple objectives which, however, have to be constructed in a well-thought-out and, if possible, empirically-based sequence. The search for elementary objectives leading to a complex one is called 'task analysis' or 'operationalisation' of the objective. Because goals expressed in terms of behaviour are necessarily of a very low level of abstraction, great attention must in practice be paid to the problem of operationalising complex goals. As a result, formulating and grouping objectives bulks large in the behaviourist-oriented development of curricula. Catalogues, tables, classifications and taxonomies are produced in order to make the task easier. Research, for its part, is mainly concerned with extending and refining the instruments of behavioural control. The most consistent and obvious applications of this stimulus–response design are, of course, programmed learning and computer-assisted instruction.

It is the perfect organisation of the content that accounts for the appeal of this approach. It is a significant aid for the teacher, the student and the school administration. The unfolding of the teaching process in individual steps from well-defined starting points (i.e. from previous sub-goals) towards precise targets, provides the teacher with a means for controlling his pupil's achievement. It also helps him to control his own work and make any necessary corrections. Pupils are provided with (in theory) lucid and unambiguous directions and with a clear control system. The administration has no trouble selecting the material.

The approach itself is, of course, content-neutral: it must be given suitable content in order to furnish a concrete curriculum. In this respect traditional mathematics was bound to appear heaven-sent, for it had anticipated the need to divide content into tasks and exercises with provision for the immediate evaluation of single steps. However, the content of the 'New Math' seemed no less compatible with the behaviourist approach; it even supplied its own clear systematisation and precise formulation of its organisation.

Case studies

The extracts below* illustrate the writings and methods of Gagné, a leader of the behaviourist school. The first extract is taken from the second edition (1970) of his influential book *The Conditions of Learning*. In the third edition (1977) Gagné demonstrates a marked change of heart; he does not attempt so clear cut a classification and no longer lists the 'eight types'.

The second example, first published in 1962, attempts to provide a hierarchical breakdown of the 'tasks' of adding integers, and constructing 'a logical demonstration of the addition of integers'. This learning hierarchy was then used by UMMaP in the development of programmed-instruction booklets for seventh-graders.

* Reproduced from The Conditions of Learning, Second Edition, by Robert M. Gagné. Copyright 1965, 1970, by Holt, Rinehart and Winston, Inc. Reprinted by permission of Holt, Rinehart and Winston.

1.   **Learning types and learning theory.** (R. Gagné, 1965, pp. 62–6.)

Eight different classes of situation in which human beings learn have been distinguished, that is, eight sets of conditions under which changes in capabilities of the human learner are brought about. The implication is that there are eight corresponding kinds of changes in the nervous system which need to be identified and ultimately accounted for. Each of these may involve different initial states or different structures, or both. From the standpoint of the outside of the human organism, however, they seem to be clearly distinguishable from one another in terms of the conditions that must prevail for each to occur. Might there actually be seven, nine, or ten rather than eight? It is quite possible that as research is continued it will become necessary to make new formulations of these conditions, to separate some or—what appears less likely—to collapse some. The distinctions made here are simply those that appear to be consistent with present evidence, much of it based on simple observation.

In brief, the varieties of learning that can currently be distinguished are as follows:

Type 1: Signal Learning. The individual learns to make a general, diffuse response to a signal. This is the classical conditioned response of Pavlov.*

Type 2: Stimulus-Response Learning. The learner acquires a precise response to a discriminated stimulus. What is learned is a connection or a discriminated operant, sometimes called an instrumental response.

Type 3: Chaining. What is acquired is a chain of two or more stimulus-response connections. The conditions for such learning have been described by Skinner and others, notably Gilbert.

Type 4: Verbal Association. Verbal association is the learning of chains that are verbal. Basically, the conditions resemble those for other (motor) chains. However, the presence of language in the human being makes this a special type because internal links may be selected from the individual's previously learned repertoire of language.

Type 5: Discrimination Learning. The individual learns to make *n* different identifying responses to as many different stimuli, which may resemble each other in physical appearance to a greater or lesser degree. Although the learning of each stimulus-response connection is a simple type 2 occurrence, the connections tend to interfere with each other's retention.

Type 6: Concept Learning. The learner acquires a capability of making a common response to a class of stimuli that may differ from each other widely in physical appearance. He is able to make a response that identifies an entire class of objects or events. Other concepts are acquired *by definition*, and consequently have the formal characteristics of rules.

Type 7: Rule Learning. In simplest terms, a rule is a chain of two or more

* Detailed bibliographic references, omitted in this extract, are provided by
   Gagné.

concepts. It functions to control behavior in the manner suggested by a verbalized rule of the form, 'If *A*, then *B*', where *A* and *B* are previously learned concepts. However, it must be carefully distinguished from the mere verbal sequence, 'If *A*, then *B*', which, of course, may also be learned as type 4.

Type 8: Problem Solving. Problem solving is a kind of learning that requires the internal events usually called thinking. Two or more previously acquired rules are somehow combined to produce a new capability that can be shown to depend on a 'higher-order' rule.

To distinguish a number of varieties of learning is not, of course, a completely novel idea. Learning types have been distinguished by many other writers. One of the most widely accepted distinctions is that between the classical conditioned response (here called type 1) and trial-and-error learning (type 2). Thorndike believed that this was a valid distinction, and Skinner considers it a basic and essential one. Hull considered the distinction between these two learning types to be one of different experimental conditions rather than of different underlying mechanisms. Several learning theorists have further distinguished chaining (type 3) as a separately identifiable form of learning; these include Skinner and also Hull in his treatment of habit-families hierarchies. Some investigators of learning describe a number of different types of distinctive learning; Tolman distinguished six kinds; Woodworth, five. Modern writers have tended to pay increasing attention to the more complex forms of learning. For example, Mowrer discusses discriminations and concept learning (type 6), as well as the simpler varieties. Harlow has studied the acquisition of concepts in comparison with simpler discrimination in monkeys. A detailed analysis would reveal many ideas within the present volume that are derived from these writers.

Prerequisites to Learning

Throughout many years of experimental investigation of learning, there have been those who have contended that *all learning is basically the same.* Thorndike, for example, says essentially this, and there have been many others who have espoused this view, explicitly or implicitly. It should be perfectly clear from the present chapter that it is this viewpoint about learning which is categorically rejected. The attempt is made to show that each variety of learning described here begins with a *different state of the organism* and ends with a *different capability for performance.* It is believed, therefore, that the differences among these varieties of learning far outweigh their similarities. Furthermore, great confusion can arise – and has arisen – through believing that these varieties are somehow alike. To equate the responding of an animal to a warning signal with the learning of a child asking for a doll, or the learning of a student to identify a chromosome or the learning to predict inheritance with the laws of genetics is considered to be a matter of gross disregard for some obvious and simple observations.

The most important class of conditions that distinguishes one form of learning from another is the initial state of the learning – in other words, its *prerequisites*. The conditions for chaining, for example, require that the individual has previously learned stimulus–response connections available to him, so that they *can* be chained. If this condition is not met, one finds oneself dealing with conditions for establishing these prerequisite $Ss \to R$'s, and thus one is likely to draw incorrect conclusions about chaining itself. This generalization, applied to the varieties of learning we have discussed, may be briefly stated as follows:

Problem Solving (Type 8)
|
requires as prerequisites:
|
Rules (Type 7)
|
which require as prerequisites:
|
Concepts (Type 6)
|
which require as prerequisites:
|
Discriminations (Type 5)
|
which require as prerequisites:

Verbal associations (Type 4)

or other Chains (Type 3)   or

which require as prerequisites:

Stimulus-Response connections (Type 2)

It is tempting to agree with Mowrer that $Ss \to R$ connections (type 2) require signal learning (type 1) as a prerequisite. This may be true, but it does not seem possible to draw this conclusion with complete confidence from presently available evidence; it remains as a proposition to be further illuminated by experimental research.

2.   **A specimen 'learning hierarchy'**

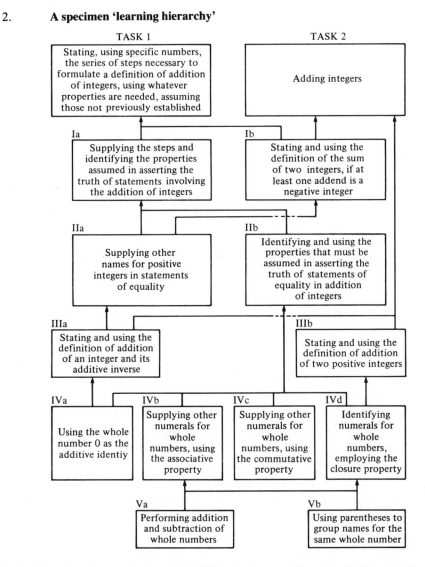

A learning hierarchy pertaining to the addition of integers. (From R. M. Gagné, J. R. Mayor, H. L. Garstens, and N. E. Paradise, Factors in acquiring knowledge of a mathematical task. *Psychol. Monogr.*, 1962, **76,** No. 526, Figure, 1, p. 4. Copyright 1962 by the American Psychological Association.

Exercises

1 Give mathematical examples corresponding to each of Gagné's varieties of learning.
2 Compare the second and third editions of Gagné's book paying particular attention to change of emphasis.
3 Work through the 'learning hierarchy' on p. 99, explaining the 'content' of each box in a greater detail.
4 Select another topic and construct a corresponding hierarchical structure. Compare this with the treatment of the topic in a textbook series. Are there noticeable omissions or differences of approach in the textbook treatment?
5 Study a typical behaviourist taxonomy, e.g. Scandura's *Mathematics: concrete behavioural objectives* (1971). Compare this with other taxonomies, for example, Wood (1968), Manheim *et al.* (see Bloom, *et al.*, 1971), Avital & Shettleworth (1968).
6 It has been asserted that behaviourists concentrate on predictable educational goals, whereas some of the most important goals of education are non-predictable. Discuss and, if necessary, exemplify this assertion.

3.2. *The new-math approach*

This approach was an indirect by-product of the work of the Bourbaki group, who had offered in their treatise a systematic description of mathematics, reorganised so as to emphasise structural considerations and presented in a uniform language with great precision. Because of its clarity and the impressiveness with which it gradually unfolds, the structural presentation offers a convincing (or, perhaps, tempting) way of organising and presenting the teaching of mathematics in the universities. By comparison, the content of school mathematics seemed to lack both accuracy and systematic organisation. A stimulus was, therefore, provided to reform traditional school mathematics, which clearly was neither successful, nor in accord with prevailing scientific standards. The resulting reorganisation resulted in significant changes in the selection and mathematical treatment of content. In the elementary schools, content, mainly number work, did not greatly change, but the way in which it was developed – from set theory – and the emphasis laid on structural aspects, for example, the commutative, associative and distributive laws, did. Certain traditional secondary-school topics, such as Euclidean geometry and advanced trigonometry, were discarded as having no significant function in the new system and in their place came probability, statistics and some computer science. The basic principle of Bourbaki, the deduction of content from axioms, also became central to the teaching of mathematics. Content which was not amenable to a deductive approach, such as many of the applications of mathematics, was relegated to the background: the ability to supply a mathematical proof and to argue logically was considered more important than the acquisition of 'trivial computing skills'. (Non-trivial skills were, unfortunately, also often to go by default.) As a result of the reorganisation it became possible to treat highly demanding domains of mathematics at early stages of the child's school life without having to make concessions, untenable from the scientific point of view, to his understanding. The approach, however, did not say anything about how such basic concepts could be introduced at the elementary level in a manner consistent with the systematic organisation of the material. Moreover, it was aimed almost exclusively at the better preparation of those who would go on to study mathematics at university. In contrast to the behaviourist approach, it saw the reform of the mathematics curriculum in terms of a renewal of content: it made demands concerning methods and did not question existing teaching practice. Rather it was based on the transmission of content through lectures and instruction. This was made clear in its provisions for pre- and in-service education which were exclusively subject-matter oriented.*

Case study

Dieudonné's address to the OEEC seminar held at Royaumont in 1959 is a landmark in the history of modern school mathematics. We give below an edited version (OEEC, 1961, pp. 31–45, reproduced with permission of the publishers) which illustrates the 'New-Math' conception and also gives a good impression of Dieudonné's colourful and forceful style.

New Thinking in School Mathematics

My specific task today is to examine, from the point of view of the present curriculum in mathematics in universities and engineering schools:

a) What mathematical background professors in these institutions would like to find in the students at the end of their secondary school years.

---

* Readers in Britain may find some of the remarks above applicable to a number of British projects. Yet the full force and 'purity' of the 'New-Math' approach was never encountered in Britain as it was, say, in the USA, France and Belgium. The only close comparison would be that provided by the 'Swansea Scheme'.

b) What they actually get.

c) How it would be possible to improve the existing situation.

. . .

In the last 50 years, mathematicians have been led to introduce not only new concepts but a new language, a language which grew empirically from the needs of mathematical research and whose ability to express mathematical statements concisely and precisely has repeatedly been tested and has won universal approval.

But until now the introduction of this new terminology has (at least in France) been steadfastly resisted by the secondary schools, which desperately cling to an obsolete and inadequate language. And so when a student enters the university, he will most probably never have heard such common mathematical words as set, mapping, group, vector space, etc. No wonder he is baffled and discouraged by his contact with higher mathematics.

Some elements of calculus, vector algebra and a little analytic geometry have recently been introduced for the last two or three years of secondary school. But such topics have always been relegated to a subordinate position, the centre of interest remaining as before 'pure geometry taught more or less according to Euclid, with a little algebra and number theory'.

I think the day of such patchwork is over, and we are now committed to a much deeper reform – unless we are willing to let the situation deteriorate to the point where it will seriously impede further scientific progress. And if the whole program I have in mind had to be summarised in one slogan it would be: *Euclid must go!*

This statement may perhaps shock some of you, but I would like to show you in some detail the strong arguments in its favor. . . . It is thanks to the Greeks that we have been able to erect the towering structure of modern science.

But in so doing, the basic notions of geometry itself have been deeply scrutinised, especially since the middle of the 19th century. This has made it possible to reorganise the Euclidean *corpus,* putting it on simple and sound foundations, and to re-evaluate its importance with regard to modern mathematics – separating what is fundamental from a chaotic heap of results with no significance except as scattered relics of clumsy methods or an obsolete approach.

The result may perhaps be a bit startling. Let us assume for the sake of argument that one had to teach plane Euclidean geometry to mature minds from another world who had never heard of it, or having only in view its possible applications to modern research. Then the whole course might, I think, be tackled in two or three hours – one of them being occupied by the description of the axiom system, one by its useful consequences and possibly a third one by a few mildly interesting exercises.

Everything else which now fills volumes of 'elementary geometry' – and by that I mean, for instance, *everything* about triangles (it is perfectly feasible and desirable to describe the whole theory without even *defining* a triangle!) . . . – has just as much relevance to what mathematicians (pure and applied) are doing today as magic squares or chess problems!

If this seems fantastic, let me go into a few details. (*Here Dieudonné describes (A) the axioms for a two-dimensional real linear space and (B) those for an inner product.*)

What I called the useful consequences are, on one hand, two-dimensional linear algebra (linear dependence, bases, straight lines, the group of translations and homothetic mappings, parallel lines, linear mappings, linear forms and equations of lines), which is exclusively derived from the system of axioms *(A),* and constitutes what is also called plane *affine geometry,* and, on the other hand, orthogonality, circles, rotations, symmetries, angles and the group of isometries, which stem from *(B).*

Of course, from this point of view, the old quarrel of 'pure' versus 'analytic' geometry becomes meaningless, both being mere translations of the language of vectors (which, by the way, it is often better to apply directly). It is clear how three-dimensional geometry can be developed exactly along the same lines. . . .

In contrast with this 'ideal' way of teaching geometry, I do not have to tell you what is actually done at present in the secondary schools. The basic notions (point, line, distance, angle) are never given a strict axiomatic definition; they are introduced by direct appeal to intuition, although their exact relation to the physical objects they are supposed to 'idealise' is never made very clear. As no complete system of axioms is ever stated, it is, of course, completely impossible to check the correctness of any proof.

The defenders of 'tradition-at-any-cost' have, of course, a ready-made answer to this. If one is to believe them, Euclidean geometry, as taught in their way, is the *only* method by which the child's mind may be opened to a real understanding of mathematics. However, as no other approach has ever been tried, I hardly see how this claim can be taken, other than as an article of faith.

Furthermore, they will add that, after all, the great mathematicians of the past and of today have been educated that way and it did not prevent them from making their discoveries. This is certainly true, but I am convinced that if these mathematicians had been taught *nothing at all* until the age of, say, 16, they would, in all probability, have done just as well, I may add, moreover, that these usual arguments carry very little weight. *Nobody need be concerned, in secondary schools at least, with teaching the future professional mathematicians* (not to speak of the great ones), of which there may be one in 10 000 children. What is really at stake is the kind of mental picture of mathematics which will emerge in the mind of an *average* intelligent student after he has been subjected to that treatment for several years.

. . .

If we had a curriculum at last freed from the deadweight of 'pure geometry', what would we put in its place? I have already mentioned briefly some of the topics which would comprise an extremely valuable preparation for higher level theories; in more detail, I would list the following ones:

a) Matrices and determinants of order 2 and 3.
b) Elementary calculus (functions of *one* variable).
c) Construction of the graph of a function and of a curve given in parametric form (using derivatives).
d) Elementary properties of complex numbers.
e) Polar co-ordinates.

I claim that none of these involves more abstract or deeper thinking than classical geometry, provided their teaching is adapted to the intellectual development of the students. This means, of course, that there remain several big problems, the main one being how to organize this material into a well-balanced curriculum and to devise methods of teaching it.

. . .

With these ideas in mind, let us turn to an outline of what I think a modern curriculum might be. I will divide it according to the age of the students (in order to avoid national peculiarities and distribution into various grades or classes); and at each level I will discuss the 'experimental' and the 'deductive' aspects of the various topics.

Ages to 14. It is probably wise to limit the teaching of mathematics in this period to 'experimental' work with algebra and plane geometry and to make no attempt at axiomatization. This does not mean that logical inferences should not be stressed whenever it is possible to point them out in an unmistakable way.

In algebra, the goal is to make the student thoroughly familiar with the technique of computation with letters, the notion of negative number, and the working out of linear problems with one or two unknowns; this is what is essentially done at present, and I have therefore no modifications to propose here, except that I would like to see more hours spent on algebra than on geometry at that stage.

With regard to geometry, I understand that much research and experimentation has lately been going on . . . concerning the methods by which this teaching of geometry as a part of physics, so to speak, can be conducted. I think this development should be highly encouraged, provided it puts the emphasis *not* on such artificial playthings as triangles, but on basic notions such as symmetries, translations, composition of transformations, etc.

Finally, in all these 'experimental' mathematics, the language and notations now universally in use should be introduced as soon as possible: there is nothing mysterious or forbidding in abbreviating 'belong to' by $\in$, or 'implies' by $\Rightarrow$, or in speaking of 'subset' instead of 'geometrical locus'. And calling some object by its proper name – like 'group' or 'equivalence relation' whenever such an object is naturally observed in some algebraic or geometric setting – does not at all imply that one has necessarily to develop in advance the *abstract* theory of groups or of equivalence relations.

If it is thought advisable, psychologically speaking, to start some axiomatics at that level, then, according to our general principle, we should look to the part of mathematics with which the children have had the most protracted contact on the 'experimental' side, namely elementary arithmetic.

Indeed, it is one of the simplest and most beautiful exercises in logic to develop the usual laws of arithmetic starting from the 'Peano axioms', and I see no reason why this should not be attempted at the earliest possible moment.

. . .

It goes without saying that this should not be started before the student can be made to understand the need for such an axiomatic treatment by urging him to think of what is meant by very large integers and why we accept the validity of arithmetic laws for such objects completely beyond our intuitive grasp; but I think I need not stress such points, of which you are far more aware than I can possibly be.

Age 14. On the 'experimental' side, this is the age at which the idea of the graph of a function is introduced, and it should certainly not be postponed any longer. To this idea should be immediately related the general method of solving an equation $f(x) = 0$ with the help of the graph of $y = f(x)$, and the various approximation processes (Lagrange, Newton) for the numerical computation of the roots, which are derived from that idea.

The emphasis here should be on approximate solutions and *never* on so-called 'closed formulas' for the roots; the student should be warned never to expect such formulae except under extremely special circumstances. In particular, the formula for solving a quadratic equation should be barely mentioned at that stage, and no special study should be devoted to that type of equation. . . .

On the 'logical' side, it seems that at this point, after several years of algebra, the time is ripe for an axiomatic description of real numbers. By this, I do *not* mean, of course, the traditional *construction* of real numbers by Dedekind cuts or Cantor fundamental sequences starting from the rational numbers. . . .

What I have in mind is much more pedestrian (and also much more useful and illuminating). It consists solely of listing the basic properties of real numbers from which all others can be logically derived. It is well known that these properties can be summed up by saying that the real numbers form an Archimedean ordered field, in which the principle of nested intervals holds.

Nor do I propose that in secondary schools any attempt should be made at proving the more difficult theorems on real numbers, such as the existence of the maximum or Bolzano's theorem on the existence of roots (even for polynomials). However, I think it should be stated clearly that these results are provable on the basis of the axioms, intuitive as they may look.

Age 15. By this time the previous study of *plane* geometry from the 'experimental' standpoint should have prepared the student for the statement of axioms (A) and (B) as given above. The consequences of these axioms should, of course, be developed both from the algebraic and the geometric point of view, i.e. any notion should be given with both kinds of interpretation. As usual, the emphasis should be on the linear transformation, their various types and the groups they form. Matrices and determinants of order 2 appear, of course, in a natural way during this development.

When this is done, the 'experimental' study of secondary-school mathematics, properly speaking, is at an end, since all the axioms have now been formulated. But in the study of any theory there is still room for shifting the emphasis either to the technical or to the conceptual side of the notions one has to introduce. And, according to our principle, any new theory has a better chance of being assimilated through its technical aspects rather than by dwelling on fine points of logical deduction.

This applies in particular to the beginning of differential calculus for functions of *one* variable, which I think is best placed at that age. I have, therefore, no quarrel with the way in which this teaching is usually done, provided the main notions of limit and continuity have been correctly defined; it is advisable to skip the proofs of all the theorems of calculus (but not their precise statement) and to concentrate on the practical techniques of the computation of derivatives and of using them for graphing functions and solving equations.

Age 16. The axiomatic part should further develop the consequences of the axioms, with a deeper study of the groups of plane geometry, and, in particular, the use of angles and of trigonometric functions. The 'measure' of angles should be defined in a precise way (as a homomorphism of the group of real numbers onto the group of rotations) but its existence admitted without proof. With this there naturally comes the introduction of complex numbers and of their geometrical interpretation.

Finally, a further topic of interest could be the discussion of all possible quadratic forms in the plane, which is equivalent to the classification of conics.

On the 'technical' side, one could start the study of the notions of a primitive and of area for simple types of domains, with elementary examples. The students could also begin to learn how to construct curves given in parametric form.

Age 17. In this final year of secondary school, the axioms for three-dimensional geometry should finally be introduced, together with their usual consequences, including of course the use of matrices and determinants of order 3.

On the more technical side, one could explain the use of primitives for computing simple types of volumes, and introduce the notion of polar co-ordinates and the method of construction of a curve given by an equation in polar co-ordinates.

Finally, logarithms and exponentials can certainly be defined and studied at that age (without any existence proof), with the emphasis on the fact that they constitute group homomorphisms.

To round off this curriculum, let me add a few words to indicate how it could connect directly with the present programme of the first years in the university. There the main topics are:

a) Linear algebra in its general form (vector spaces of arbitrary dimension, general theory of matrices and determinants).
b) Quadratic forms and finite dimensional Euclidean spaces.
c) Derivatives and integrals of functions of several real variables, with their various applications. Differential and partial differential equations. Elementary differential geometry.
d) Elementary theory of metric spaces, Banach spaces, Hilbert space and other functional spaces. Elementary functional analysis.

Exercises

1 Write an essay on the place of the triangle in school mathematics.
2 'Nobody need be concerned, in secondary schools at least, with teaching the future professional mathematicians of which there may be one in 10 000 children'. Discuss this statement and its implications concerning what is meant by a 'professional mathematician'.
3 Discuss the problems on the implementation level which might have been expected to arise from any attempt to realise Dieudonné's proposals.
4 The introduction of 'New-Math' curricula met some criticism, e.g. in the memorandum 'On the Mathematics curriculum of the high school' (Ahlfors and 51 other American and Canadian mathematics professors, 1962) and Kline (1961, 1973). Try to evaluate these criticisms in the light of Dieudonné's intentions.
5 'From a purely technical point of view . . . we are under a continuing obligation to all our . . . students to eliminate dead, useless, outmoded or unimportant parts of mathematics . . . and to offer them the liveliest and most stimulating aspects of mathematics in order to develop their understanding and their creative talents.' This statement is by Marshal Stone, chairman of the Royaumont conference. Discuss its implications relating to teaching methods, motivation, etc. Who decides what is 'stimulating'?

3.3. *The structuralist approach*

The structuralist approach is based on investigations conducted by genetic epistemology theorists into the processes of concept formation, from which Jerome Bruner developed his theory of the 'structures of the disciplines'. In broad outline the theory proceeds as follows: cognitive structures are combinations of acquired concepts and thinking abilities. Simple structures, made of a few concepts, are developed into more elaborate ones through the addition of new concepts. At their highest stage of development, cognitive structures correspond to the structure of the sciences, taken as the essence of all concepts and processes contained in them. These structures are so complex that they include all insights, concepts and procedures of the sciences, but they are simultaneously so easy to formulate that they can be transmitted at a lower level of cognition. The purpose of the transmission of the structures of the scientific disciplines is not the acquisition of the knowledge of these structures by pupils; it is not so much a matter of transmitting in an elementary manner the structures as a content of education, as of displaying their process character. This, to a large extent, is the basis for the correspondence between scientific and cognitive structures, and a means of promoting cognitive processes in pupils. Operating within the structures of the sciences reinforces the processes by which new concepts are assimilated by a given stage of cognitive development. The learner is thus given the opportunity to gain familiarity with the structures and to obtain a better grasp of their complexity which, in turn, will further the acquisition of new concepts, etc., until a complete correspondence between the learner's cognitive structures and those of the science is achieved, when the learner will have become a scientist. Bruner employs the concept of the 'spiral curriculum' to refer to this means of progress. From his theory he derives the claim that the structures of the sciences are suitable for promoting learning processes in an optimal way, thus justifying *a posteriori* the efforts made to orientate curricular reform to the structure of the scientific discipline.

The critical point of this theory is how to transmit these scientific structures to pupils endowed with lower cognitive structures. Bruner assumes that they can also be understood as generalisations and abstractions, even if only part of what is summarised and generalised can be perceived. He states further that the learner is capable of anticipating the stage of his own knowledge by grasping facts intuitively. Finally, the manner in which these goals are to be achieved is through 'discovery learning', which is simultaneously a means and an end.

Whilst 'discovering', the exploring student will acquire structures and, in so doing, behave like a scientist having full mastery of the structures. The spiral curriculum will ensure that progression takes place from lower to higher levels. At the lower levels of cognition mathematical objects are discovered empirically in the pupil's environment. Later on, analytic thinking is developed step by step, and the axiomatic method is used for explaining and analysing mathematical structures. Again, the spiral model ensures that the same objects are periodically treated anew at a higher cognitive level. The mathematical objects themselves only function as examples and rank lower than processes, methods and working procedures. The conception of science on which school mathematics is based is not the finished edifice of mathematics, but its gradual construction with emphasis placed on the processes of its organisation. The major task for the curriculum developer is to devise appealing and meaningful teaching models for these processes of discovery as embodiments of the underlying structures.

Case studies

The intentions of the structuralist approach are illustrated by two brief extracts from the *Report of the 1963 Cambridge Conference on School Mathematics* (see p. 153). This provides evidence of what has been called the 'cognitive turn', i.e. the transition from the New-Math approach to the structuralist one.

We then quote from the most eminent representatives of the structuralist school, J. S. Bruner and Z. P. Dienes. The Bruner extract is taken from a paper 'On learning mathematics' first published in *The Mathematics Teacher,* 1960; that by Dienes summarises the thesis and content of his book *The Six Stages in the Process of Learning Mathematics* (1973), reproduced by permission of the National Foundation for Educational Research in England and Wales.

1. *Goals for School Mathematics* (Cambridge Conference, 1963, *pp. 10, 17.)*
    The Role of 'Modern' Mathematics
    Some have argued that the mathematics curriculum should be organized to provide the quickest possible introduction to contemporary mathematical research. This view we reject. Contemporary mathematical research has given us many new concepts with which to organize our mathematical thinking; it is typical of the subject that some of the most important of these are very simple. Concepts like set, function, transformation group, and isomorphism can be introduced in rudimentary form to very young children, and repeatedly applied until a sophisticated comprehension is built up. We believe that these concepts belong in the curriculum not because they are modern but because they are useful in organizing the material we want to present. . . .

    Discovery approach. The discovery approach, in which the student is asked to explore a situation in his own way, is invaluable in developing creative and independent thinking in the individual. In this system memorizing a mechanical response does not help the student to advance. His innate interest and competitive nature force him to concentrate on the creative problem at hand.

It is obvious, however, that the discovery method is slow. It took mankind thousands of years to discover, collectively, the concepts we wish to teach. Cut off from communication with the knowledge of others, the student can proceed but a little way along the path of wisdom in his allotted time. This is at least tacitly admitted by all proponents of discovery, but there is considerable variation in the amount of feedback and reinforcement used to guide the student. As a minimum the context and the very statement of the problem, or the equipment given to work with, is a guide to the student – a very important one.

We believe that usually one should go farther than this in aiding discovery: that the teacher should be prepared to introduce required ideas when they are not forthcoming from the class; that he should bring attention to misleading statements in the way of the discussion, and summarize results clearly as they come forward. He should not allow the 'moments of triumph' to pass by unnoticed.

2.   *On learning mathematics* (J. S. Bruner, 1960, *The Mathematics Teacher* **53,** pp. 68–70.)

How can we state things in such a way that ideas can be understood and converted into mathematical expression?

It seems to me there are three problems here. Let me label them the *problem of structure,* the *problem of sequence,* and the *problem of embodiment.* When we try to get a child to understand a concept, leaving aside now the question of whether he can 'say' it, the first and most important problem, obviously, is that we as expositors understand it ourselves. I apologize for making such a banal point, but I must do so, for I think that its implications are not well understood. To understand something well is to sense wherein it is simple, wherein it is an instance of a simpler, general case. I know that there are instances in the development of knowledge in which this may not prove to be the case, as in physics before Mendeleev's table or in contemporary physics where particle theory is for the moment seemingly moving toward divergence rather than convergence of principles. In the main, however, to understand something is to sense the simpler structure that underlies a range of instances, and this is notably true in mathematics.

In seeking to transmit our understanding of such structure to another person – be he a student or someone else – there is the problem of finding the language and ideas that the other person would be able to use if he were attempting to explain the same thing. If we are lucky, it may turn out that the language we would use would be within the grasp of the person we are teaching. This is not, alas, always the case. We may then be faced with the problem of finding a homologue that will contain our own idea moderately well and get it across to the auditor without too much loss of precision, or at least in a form that will permit us to communicate further at a later time.

Let me provide an example. We wish to get across to the first-grade student that much of what we speak of as knowledge in science is indirect, that we talk about such things as pressure or chemical bonds or neural inhibition although we never encounter them directly. They are inferences we draw from certain regularities in our observations. This is all very familiar to us. It is an idea with a simple structure but with complicated implications. To a young student who is used to thinking of things that either exist or do not exist, it is hard to tell the truth in answer to his question of whether pressure 'really' exists. We wish to transmit the idea that there are observables that have regularities and constructs that are used for conserving and representing these regularities, that both, in different senses, 'exist', and the constructs are not fantasies like gremlins or fairies. That is the structure.

Now there is a sequence. How do we get the child to progress from his present two-value logic of things that exist and things that do not exist to a more subtle grasp of the matter? Take an example from the work of Inhelder and Piaget. They find that there are necessary sequences or steps in the mastery of a concept. In order for a child to understand the idea of serial ordering, he must first have a firm grasp on the idea of comparison – that one thing includes another or is larger than another. Or, in order for a child to grasp the idea that the angle of incidence is equal to the angle of reflection, he must first grasp the idea that for any angle at which a ball approaches a wall, there is a corresponding unique angle by which it departs. Until he grasps this idea, there is no point in talking about the two angles being equal or bearing any particular relationship to each other, just as it is a waste to try to explain transitivity to a child who does not yet have a firm grasp on serial ordering.

The problem of embodiment then arises: How to embody illustratively the middle possibility of something that does not quite exist as a clear and observable datum? Well, one group of chemists working on a new curriculum proposed as a transitional step in the sequence that the child be given a taped box containing an unidentified object. He may do anything he likes to the box: shake it, run wires through it, boil it, anything but open it. What does he make of it? I have no idea whether this gadget will indeed get the child to the point where he can then more easily make the distinction between constructs and data. But the attempt is illustrative and interesting. It is a nice illustration of how one seeks to translate a concept (in this instance the chemical bond) into a simpler homologue, an invisible object whose existence depended upon indirect information, by the use of an embodiment. From there one can go on.

The discussion leads me immediately to two practical points about teaching and curriculum design. The first has to do with the sequence of a curriculum, the second with gadgetry. I noted with pleasure in the introductory essay of the Twenty-fourth Yearbook of the National Council of Teachers of Mathematics that great emphasis was placed upon continuity of understanding:

Theorem 2. Teachers in all grades should view their task in the light of the idea that the understanding of mathematics is a continuum. . . . This theorem implies immediately the corollaries that: (1) Teachers should find what ideas have been presented earlier and deliberately use them as much as possible for the teaching of new ideas. (2) Teachers should look to the future and teach some concepts and understandings even if complete mastery cannot be expected.

Alas, it has been a rarity to find such a structure in the curriculum, although the situation is likely to be remedied in a much shorter time than might have been expected through the work of such organizations as the School Mathematics Study Group. More frequently fragments are found here and there: a brilliant idea about teaching co-ordinate systems and graphing, or what not. I have had occasion to look at the list of teaching projects submitted to the National Science Foundation. There is everything from a demonstrational wind tunnel to little Van de Graaff generators, virtually all divorced from any sequence. Our impulse is toward gadgetry. The need instead is for something approximating a spiral curriculum, in which ideas are presented in homologue form, returned to later with more precision and power, and further developed and expanded until, in the end, the student has a sense of mastery over at least some body of knowledge.

There is one part of the picture in the building of mathematical curriculum now in progress where I see a virtual blank. It has to do with the investigation of the language and concepts that children of various ages use in attempting intuitively to grasp different concepts and sequences in mathematics. This is the language into which mathematics will have to be translated while the child is en route to more precise mastery. The psychologist can help in all this, it seems to me, as a handmaiden to the curriculum builder, by devising ways of bridging the gap between ideas in mathematics and the students' ways of understanding such ideas. His rewards will be rich, for he not only will be helping education toward greater effectiveness, but also will be learning afresh about learning. If I have said little to you . . . about the formal psychology of learning as it now exists in many of our university centers, it is because most of what exists has little bearing on the complex and ordered learning that you deal with in your teaching.

3.   *The Six Stages in the Process of Learning Mathematics* (Z. P. Dienes, 1973*b*, pp. 6–9.)
        First stage
        The notion of the environment seems to us to be of outstanding importance, for, in a certain sense, all learning is basically a process by which the organism adapts to its environment. To say of a child, or, in a completely general way, of any organism, that he has learnt something, signifies that the organism (or child) has been able to modify his behaviour in relation to some aspect of his environment. In the phase which precedes any learning, the

organism is badly adjusted to the environment. However, thanks to the learning, the organism becomes able to adapt himself to a point where he is capable of dominating the situations which he meets in that environment. If one accepts that it is the process of adaptation which represents all learning, it would seem reasonable to present the child with a suitable environment to which he might adapt himself if we wish learning to take place.

To be more precise, the adaptation takes place in what might be designated as the free play stage. All children's games represent a kind of exercise which helps the child to adapt to situations which he is going to meet in later life. But, if one proposes to teach logic to a child, it seems necessary to confront him with some situations which will lead him to form logical concepts. Let us consider for a moment the example of logic; we need to recognize that the natural environment in which the child lives does not embrace all the attributes which we consider as logical. It is necessary, then, to invent an artificial environment. As a consequence of this environment the child will be led little by little to form logical concepts in a more or less systematic way. An example of such an environment might comprise, for example, the universe of logic blocks. . . . If one wanted the child to make his first steps towards learning ideas concerned with the concept of powers, it is advisable that he be put in the corresponding environment. Such an environment might be constructed with multibase materials. . . . It is possible to give a large number of similar examples to show how one might create a special environment for the learning of any set of mathematical ideas whatsoever.

Second stage

After a certain period of adaptation, that is free play, the child will realize some constraints in the situations. There are certain things which one may not do. There are certain conditions which must be satisfied before certain goals can be attained. The child will soon realise that there are some regularities in the situation. At this moment, he will be ready to play with the restrictions which have been artificially imposed. These restrictions may be called the rules of the game.

When one plays chess, it is completely arbitrary that certain pieces have certain properties in the game. These properties are in no way dependent upon the shape or any other physical property of the pieces. In just the same way, one may suggest to the children games with some rules. Later the children themselves may be able to invent other rules, or change the rules and play the corresponding games. Thus they become used to the manipulation of these regularities. Clearly, if one wishes the child to learn about mathematical structures, the sets of rules that one suggests will depend upon the relevant mathematical structures. These games are constructed from structural materials like those described above.

Third stage

Obviously, to play structural games according to mathematical laws intrinsic in any mathematical structure is not *learning mathematics.* How is the child to be able to extract from this set of games the underlying mathematical abstractions? The psychological means of doing this is to play some games which possess the same structure, but which appear very different to the child. Thus, the child will come to realize the links of an abstract nature that exist between the elements of two apparently different games which have identical structures. This is what we call the *dictionary game,* or, if one wishes to use a more mathematical description, the *isomorphism game.* Thus the child separates out the common structure of the two games and is able to discard the non-relevant parts. . . . Thus, the games played with one embodiment, then again with another embodiment, will be identified from the point of view of their structures. It is at this moment that the child realizes what is the 'same' in the different games that he has experienced, that he will have made an 'abstraction'.

Fourth stage

Indeed, the child will not yet be in a position to make use of this abstraction, because it is not yet properly fixed in his mind. Before he can become fully aware of an abstraction, the child needs a method of representation. Such a representation allows him to talk about what he has abstracted, to look at it from outside, to dispense with the actual game or set of games, but to examine the games in general and reflect upon their subject. Such a representation might be a set of graphs, it might be a Cartesian system, a Venn diagram, or any other visual representation – or even auditory representation in the case of children who do not think in an essentially visual way.

Fifth stage

Following the introduction of a representation, or even several representations of the same structure, it will be possible to examine this representation. The goal of this examination is to understand some properties of the abstraction which has been made. In a representation, one can easily see that some principal properties of it are mathematical ones that we wished to create. This indicates that it is necessary, at the present stage, to have a description of what we have represented. For a description we must have a language. This is why the realization of the properties of the abstraction in this fifth stage must be accompanied by the invention of a language and then the use of this newly invented language to describe the representation. It will be better, if possible, for each child to invent his own language and then later the children, with the help of their teacher, can discuss among themselves whether one of these languages has more advantages than the others. Such a description forms the basis for a system of axioms. Each part of the description may serve as an axiom or later even as a theorem.

Sixth stage

Most mathematical structures are so complex that they possess an infinite number of properties. It is impossible to give all these properties in a description of a system which one has produced. It is necessary, then, by some means, to enclose the description in a finite domain, and to use a finite number of words. That is, we need a method by which we can reach any part of the possible description, given a first part as a starting point. These methods of arriving at other parts of the description will be our *'rules for proving games'*. These later descriptions which come by this method will be called the *'theorems of the system'*. Thus we have invented a formal system in which there are some *'axioms'*, that is the first part of the description, and some *'rules of play'*. There may be others which will be the *'logico-mathematical rules'* of proof. Finally, there will be the theorems of the system which are the parts reached from the initial description by using the 'rules of play'.

Exercises

1 'To understand something well is to sense wherein it is simple.' Discuss and exemplify.
2 Give an example where one can use a homologue to transmit an idea.
3 It is noticeable that in an article for mathematics teachers, titled 'On learning mathematics', Bruner draws his examples from physics and chemistry. Provide alternative examples drawn from mathematics.
4 'Fourth grade students can go a long way into "set theory".' In what sense is this true? What benefits will ensue? An important quantifier is missing – supply it.
5 Investigate the problems of preparing a 'teaching language'. (The bibliography by Austin & Howson (1979) may help.)
6 Suggest ways (e.g. through games, flowcharts), in which the idea of 'isomorphism can be introduced in a rudimentary form' and then can be reinforced until 'a sophisticated comprehension is built up'.
7 What do you think constitutes a 'sense of mastery'? Is this different from 'mastery' and, if so, in what way? Are there occasions when one should teach for 'a sense of mastery' rather than 'mastery'?
8 Discuss, criticise and, where possible, further exemplify the six stages of learning described by Dienes.
9 Do you see Dienes' stages following on at regular chronological intervals, or are there cases where successive stages would seem some considerable distance apart in a child's cognitive development?
10 Compare and contrast Gagné's types (p. 96) with Dienes' stages. What do you see as the strengths and weaknesses of the two approaches?

3.4.   *The formative approach*

The formative approach is formulated without reference to particular school subjects. It proceeds on two assumptions: first, that any school education aims at endowing the pupil with an optimal basic body of cognitive abilities and affective and motivational attitudes; second, that these factors may be described in terms of personality traits. These traits will include such factors as creative abilities, 'intelligence', and performance motivation which will be identified and described by psychologists, psychometricians, sociologists and others. The scientific analysis of such factors, or abilities, will permit their growth through careful planning, the methods and content that initiate these boosting processes being determined by the structures of personality development and not by those of the sciences.

The task of the curriculum developer, therefore, in this framework is to find out, and match adequately, the content and methods most likely to develop the desired factors or to enhance their development. The individual sciences are relevant in as much as they can contribute to and be integrated within these educational processes. In this respect, Piaget's research has demonstrated the vital contribution to be made by mathematics. By investigating the growth of mathematical concepts in children, Piaget came to realise that such concepts are formed through the internalisation of schemata of activity in the manipulation of concrete objects and their abstraction. He calls the various stages of abstraction, from concrete to logical operations, the 'levels of operative intelligence'. Because Piaget's research and that of others have been mostly related to the earlier levels of the development of intelligence, and the corresponding theories are well-elaborated for these levels, this approach has been mainly employed by projects working in elementary schools.

It is fundamental to the processes of concept formation studied by Piaget that they should be initiated at the level of concrete operations by the child's self-reliant manipulation of real objects. It does not do to represent the action symbolically by, for example, reconstructing the processes observed in investigations in terms of specially created materials: real objects must elicit the child's action in real situations. As in the structuralist concept, the necessity arises, therefore, to devise suitable initial teaching situations. The difference here, however, is the emphasis placed on reality and the rejection of models. Open-endedness becomes a basic characteristic of curricular units which then serve to initiate learning processes but not to determine them. Since one cannot ensure what will emerge from the autonomous activity of the pupil, progress of the teaching process is to some extent uncertain. For that reason, projects cannot aim at producing materials in the form of ready-to-use teaching units. Instead the products are intended to help the teacher to perform his key function, which is to create situations in which the child can enjoy 'real activity' and translate the activity into learning processes. To do this, the teacher has to be provided with a body of information and ideas, and with illustrative material which gives a paradigmatic representation of possible procedures and offers at the same time a broad choice of suggestions, to which the teacher can resort according to the particular teaching situation. Materials for pupils serve merely as aids towards the mastery of situations, and as reinforcement in the learning process at the stage where experience gained in activity is 'processed' and assimilated.

There is less emphasis on producing materials within a particular science or discipline, and where the rival claims of scientific discipline and approach are in opposition, the latter prevails. So mathematical applications, discarded by followers of the New-Math movement because they did not fit neatly into the system, now assume a greater significance, being the manifestation of mathematics in reality. Demarcation of areas of subject matter is ignored to the benefit of an over-riding emphasis on cognitive processes.

Case study

We reprint extracts from Piaget's *Science of Education and the Psychology of the Child* (1965) in the translation by Coltman (1971).* It must be remembered that Piaget himself never set out a 'Piagetian method of education', but rather sought to guide educators through his exploration of the role of activity in the growth of intelligence, and of the development of the child's thinking with respect to such universals as space, time and causality, and through his attempts to identify and describe logico-mathematical structures.

The didactics of mathematics

The teaching of mathematics has always presented a somewhat paradoxical problem. There exists, in fact, a certain category of students, otherwise quite intelligent and even capable of demonstrating above average intelligence in other fields, who always fail, more or less systematically, in mathematics. Yet mathematics constitutes a direct extension of logic itself, so much so that it is actually impossible to draw a firm line of demarcation between these two fields (and this remains true whatever interpretation we give to the

* Reproduced by permission of the publisher, Longman.

relationship: identity, progressive construction, etc.). So that it is difficult to conceive how students who are well endowed when it comes to the elaboration and utilization of the spontaneous logico-mathematical structures of intelligence can find themselves handicapped in the comprehension of a branch of teaching that bears exclusively upon what is to be derived from such structures. Such students do exist, however, and with them the problem.

It is usually answered in a rather facile way by talk about mathematical 'aptitude', . . . But, if what we have just posited as to the relationship of this form of knowledge with the fundamental operational structures of thought is true, then either this 'aptitude' . . . is indistinguishable from intelligence itself, which is not thought to be the case, or else it is related entirely, not to mathematics as such, but to the way in which mathematics is taught. In fact, the operational structures of the intelligence, although they are of a logico-mathematical nature, are not present in children's minds as conscious structures: they are structures of actions or operations, which certainly direct the child's reasoning but do not constitute an object of reflection on its part (just as one can sing in tune without being obliged to construct a theory of singing, and even without being able to read music). The teaching of mathematics, on the other hand, specifically requires the student to reflect consciously on these structures, though it does so by means of a technical language comprising a very particular form of symbolism, and demanding a greater or lesser degree of abstraction. So the so-called aptitude for mathematics may very well be a function of the student's comprehension of that language itself, as opposed to that of the structures it describes, or else of the speed of the abstraction process insofar as it is linked with such a symbolism rather than with reflection upon structures that are in other respects natural. Moreover, since everything is interconnected in an entirely deductive discipline, failure or lack of comprehension where any single link in the chain is concerned entails an increasing difficulty in following the succeeding links, so that the student who has failed to adapt at any point is unable to understand what follows and becomes increasingly doubtful of his ability: emotional complexes, often strengthened by those around him, then arise to complete the block that has been formed in an initiation that could have been quite different.

In a word, the central problem of mathematical teaching is that of the reciprocal adjustment between the spontaneous operational structures proper to the intelligence and the program or methods relating to the particular branches of mathematics being taught. This problem, in fact, has been profoundly modified during recent decades due to the transformations that have taken place in mathematics itself: by a process that is at first sight paradoxical, though in fact psychologically natural and clearly explicable, the most abstract and general structures of contemporary mathematics are much more closely linked to the natural operational structures of the intelligence and of thought than were the particular structures that provided the framework for classical mathematics and teaching methods. . . . The intelligence, however, works out and employs these structures without becoming aware of them in any consciously reflective form, not in the sense that M. Jourdain spoke prose without knowing it, but rather in the sense that any adult who is not a logician nevertheless manipulates implications, disjunctions, etc., without having the slightest idea of the way in which symbolic or algebraic logic succeeds in expressing these operations in abstract and algebraic formulas. The pedagogic problem, therefore, despite the progress realized in principle by this return to the natural roots of the operational structures, still subsists in its entirety: that of finding the most adequate methods for bridging the transition between these natural but nonreflective structures to conscious reflection upon such structures and to a theoretical formulation of them.

And it is at this point, in fact, that we once more meet the conflict of which we spoke at the beginning of this section between the operational manipulation of structures and the symbolic language making it possible to express them. The most general structures of modern mathematics are at the same time the most abstract as well, whereas those same structures are never represented in the mind of the child except in the form of concrete manipulations, either physical or verbal. The mathematician who is unaccustomed to psychology, however, may suspect any physical exercise of being an obstacle to abstraction, whereas the psychologist is used to making a very careful distinction between abstraction based on objects (the source of experiment in the physical field and foreign to mathematics) and abstraction based on actions, the source of mathematical deduction and abstraction. We must avoid believing, in fact, that a sound training in abstraction and deduction presupposes a premature use of technical language and technical symbolism alone, since mathematical abstraction is of an operational nature and develops genetically through a series of unbroken stages that have their first origin in very concrete operations. Nor must we confuse the concrete either with physical experiment, which derives its knowledge from objects and not from the actions of the child itself, or with intuitive presentations (in the sense of figurative methods), since these operations are derived from actions, not from perceptual or visually recalled configurations.

These various possible misunderstandings demonstrate that, though the introduction of modern mathematics at the most elementary stages of education constitutes a great advance in principle from the psychopedagogic point of view, the results obtained may have been, in individual cases, either excellent or questionable according to the methods employed. This is why the International Conference on Public Education (International Bureau of Education and UNESCO), at its 1956 session, inserted the following articles in its Recommendation No. 43 (The Teaching of Mathematics in Secondary Schools):

20. It is important (a) to guide the student into forming his own ideas and discovering mathematical relations and properties himself, rather than imposing ready-made adult thought upon him; (b) to make sure that he

acquires operational processes and ideas before introducing him to formalism; (c) not to entrust to automatism any operations that are not already assimilated.

21. It is indispensable (a) to make sure that the student first acquires experience of mathematical entities and relations and is only then initiated into deductive reasoning; (b) to extend the deductive construction of mathematics progressively; (c) to teach the student to pose problems, to establish data, to exploit them, and to weigh the results; (d) to give preference to the heuristic investigation of questions rather than to the doctrinal exposition of theorems.

22. It is necessary (a) to study the mistakes made by students and to see them as a means of understanding their mathematical thought; (b) to train students in the practice of personal checking and autocorrection; (c) to instil in students a sense of approximation; (d) to give priority to reflection and to reasoning, etc. (Gruber & Vonèche, 1977, pp. 701–4.)

Exercises

1 'Mathematics constitutes a direct extension of logic itself . . .' Discuss the validity of this statement. How is this view of mathematics likely to affect the construction of mathematical curricula?

2 Piaget's views on mathematics were greatly influenced by the writings of Bourbaki. (This is shown explicitly in the original passage from which we quote although that part is not reproduced here.) Investigate the assertion that those projects who were most influenced by Piaget (e.g. Nuffield, Madison) show little Bourbakian influence, whilst those which draw inspiration from Bourbaki (e.g. SSMCIS, Papy) appear never to have heard of Piaget.

3 'The teaching of mathematics . . . specifically requires the student to reflect consciously on these structures.' Discuss the importance of this statement. Does it reflect what is, or what might be? Exemplify.

4 'The central problem of mathematical teaching is that of the reciprocal adjustment between the spontaneous operational structures proper to the intelligence and the program or methods relating to the particular branches of mathematics being taught.' Study the materials of a selected curriculum development project and investigate the extent to which this 'Piagetian problem' has been solved (see, for example, Malpas, 1974).

5 What were those structures which 'provided the framework for classical mathematics and teaching methods'?

6 Piaget recognises three Bourbakian 'mother structures' of mathematics: algebraic, ordering, and topological. In particular, he

sees the last as 'the child's spontaneous form of geometry'. Discuss the validity of this claim. (See, for example, Sauvy & Sauvy (1974) and Kapadia (1974).) What are the consequences for our teaching of mathematics?

7 Give mathematical examples of implication, disjunction, the contrapositive, etc. and corresponding 'everyday' uses which could be used for exemplificatory purposes in our teaching.

8 Piaget warns against 'confusing the concrete either with physical experiment, which derives its knowledge from objects and not from the actions of the child itself, or with intuitive presentations'. Discuss, with examples, Piaget's meaning. (Further guidance can be obtained by reading Piaget's book from which this extract is taken.)

9 What are the implications of the 'formative conception' for the pre-service education of teachers? What changes would this necessitate in your country?

10 'Piagetian theory . . . is inadequate both as a psychological theory . . . and as a rationale for mathematics education.' (Lunzer, 1976.) Explain or refute.

11 Is it possible to classify the work of Caleb Gattegno as either 'structuralist' or 'formative'? If so, which description would you use? Contrast Piaget's views on apparatus with those of Gattegno (see, for example, Gattegno, 1963).

12 To what extent was/is the Association of Teachers of Mathematics (ATM) 'formative'-oriented? Which approaches are dominant in the publications *Some Lessons in Mathematics* (1964), and *Notes on Mathematics for Children* (1977)?

3.5. *The integrated-teaching approach*

This approach was developed at the same time and on the same cognitive-theoretical basis as the formative one; it seeks, however, to go beyond mere statements on methods and to consider also problems of content. The content of teaching should not be the meat supplied to plump out the skeleton of methodological requirements; it should be determined by the same intents, oriented to the development of the learner's personality, on which the methodological foundation of the approach is based. Just as the organisation of the learning processes takes into account the requirements of the cognitive development of students, so the selection of the content of learning must be geared to the interests and needs of pupils. This means that real needs and problems concerning their private and professional life, both present and future, have to be considered. Problem areas from reality will determine the content of teaching.

Such problems and problem areas can rarely be dealt with within the bounds of individual school subjects or solved using the materials, concepts and insights of a single discipline. The interaction of procedures and ideas from several disciplines will play a decisive role in any problem-solving strategy. For this reason, this approach does away with the division into subjects of instruction; instead it integrates them according to the requirements of the problem at hand. Progress in problem solving, in the sense of discovery learning, is enhanced by the application-oriented means of cognition and by the procedures of the various disciplines. In this approach the contribution of mathematics is mathematisation and the provision of models which serve to relate mathematical systems to real situations. The real context of a mathematical idea becomes the subject of the teaching and learning process.

Motivation of pupils is of great importance here. The identification of problems has to be done by the students themselves, although a minimal amount of intervention will often be necessary to provide the incentive and challenge to identify and tackle problems. Curricular units have to be flexible enough to leave open the greatest number of avenues to (and from) a problem so that the problem-solving process, and hence the progress of the learning process, can be controlled by the students themselves.

The integrated teaching approach is in some senses a rebirth of a concept that had already played a significant role in an earlier phase of US pedagogic reform, namely, project teaching. In this, similar objectives had been developed although in a manner that could best be described as philosophical-intuitive.

Case study

The *Goals of the Cambridge Conference on the Correlation of Elementary Science and Mathematics* (1969) drew attention to an integrated-teaching approach in the elementary school. We reproduce some excerpts* (pp. 7–10) in order to illustrate how the emphasis placed on mathematical topics changed: more attention was now paid to their role as aids to problem solving.

General relationship of mathematics and science to the goals

Because of the comparative simplicity with which they can go beyond the superficial, science and mathematics lend themselves to the development of attitudes of lifelong and general value. Among these are:

---

* Reprinted from *Goals for the Correlation of Elementary Science and Mathematics*, copyright 1969 by Educational Development Center, Newton, Massachusetts.

1. A healthy skepticism regarding accepted knowledge and a willingness to abandon ideas which are demonstrably erroneous.

2. The humility inherent in the realization that our understanding can never be complete, coupled with the optimism of conviction that, nevertheless, our understanding can always be increased.

3. The realization that understanding, while indeed a means to power, is a joy and an end in itself.

Two hundred years ago a scientist was likely to be conversant with all the fields of science and mathematics. Most scientists worked in only one or two fields, but they managed to keep abreast of important developments in all areas. School work at that time covered the gamut of science and mathematics under the title 'natural philosophy'. With the exponential growth of knowledge we have long since passed the day when anyone could converse knowledgeably about all fields of science; indeed, there are very few who can keep up with more than a small subarea within a field as broad as biology or mathematics. Because of this, there has been a fragmentation of natural philosophy into different disciplines – mathematics, physics, chemistry, biology, etc. With the recent wave of curricular reform, there has been very little effort to bring these subjects together.

. . .

Now that science is gaining a larger place in the elementary school, it is time to consider whether it might not be a good time to return to the concept of natural philosophy. Elementary school science has, so far, been general science, but with a few exceptions it has been almost totally divorced from mathematics. There are in fact, both science and mathematics teachers who argue that science and mathematics must be kept apart at the elementary level. They claim that the basic philosophies and points of view of the two areas are so different as absolutely to preclude treating them together. Some science teachers say that it is so hard to get children to make simple qualitative scientific observations that any move toward quantitative science is foredoomed. . . . Some mathematics teachers argue that their subject is complete within itself, and that it requires only minimal references to the outside world. They point out that children already handle rather complicated abstractions in their mathematics classes, and that to attempt to tie everything up with science, especially with experimental science, would represent a substantial retrogression.

In the opinion of a few at our conference a more serious danger is the possible disruption of a tightly organized mathematical sequence by an attempt to cater to the needs of science instruction. In this view the linear nature of mathematical logic highly restricts the order in which mathematical topics may be efficiently learned.

Recognizing that there are cogent arguments on the other side, the majority of the conference participants nevertheless took the position that a determined effort should now be made to integrate science and mathematics teaching, particularly in the elementary schools. To the argument that the basic philosophies are too divergent for a meaningful merger, we reply that the complementarity of the mathematical and the scientific points of view will provide the great strength of the combined curriculum. We would not attempt to ignore the differences between mathematics and science; rather we would emphasize that some of the greatest successes have been achieved by men who could look at problems from both points of view. To the argument that quantitative experimentation is beyond the scope of the elementary school, we point out that the work done by the Madison Project, MINNEMAST, and AAAS in this country, as well as the work of the Nuffield Foundation in England, seems to show that quantitative work can be successful. Children spend nine years in primary and junior high school. A great deal can be accomplished in such a time span if we start slowly and move ahead at a natural pace. To the argument that mathematics has stood for years on its own pedestal in the schools, we reply that it has hardly been a screaming success. For every child who feels comfortable with mathematical abstractions we can surely find two who are bewildered and repelled. Moreover, we wonder how many of those children who seem at ease with abstractions will be able to transfer their concepts to any practical situation. To the argument that the mathematical sequence would be damaged by integration with science the majority of the conferees would reply that the very use of concrete examples and new concepts and motivations emanating from the science side increases the flexibility of the organization of mathematics instruction.

An integrated mathematics-science curriculum will not be easy to achieve. It will require vastly more experimentation than has yet been done by any of the science or mathematics curriculum groups. We believe that it will require a fundamental change in the style of school instruction. We have in mind a system of semi-individualized instruction based on a large number of small units to be worked through by individual pupils or small groups. There would be many more units than any single pupil would attempt, and the principal responsibility of the teacher would lie in assigning appropriate projects to the pupils.

The merits of such a system can be argued quite apart from the question of mathematics-science integration, of course, but we think that some such system may be essential to achieve the effects we want. Obviously, there will be a major problem in the training of teachers, but this is already formidable and we doubt that it will be magnified by the integration of mathematics and science. But whatever the difficulties, it was generally agreed that the advantages of emphasizing the organic interconnections between mathematics and the various branches of science would far outweigh any disadvantages.

Although our goal of an integrated mathematics-science curriculum is a long way off, its adoption has an immediate and significant payoff. As soon as one starts thinking about an integrated curriculum for grade school, his mind is freed of the weight of the traditional arithmetic course. It becomes quite natural to ask such questions as 'Why do we spend most of the first grade on addition problems involving strictly discrete situations with almost no time devoted to addition in basically continuous contexts like lengths and weights?' 'Should a large amount of instructional time be devoted to the mastery of arithmetic algorithms?' Within an integrated framework, planning becomes a cooperative effort for scientists and mathematicians. It is no longer merely a matter of one group asking the other to modify its program to accommodate some topic the first group wants to build on.

Within the mathematics-science framework we must delineate goals consonant with our view of the whole process of education. We believe that the primary goal of science education is an understanding of scientific methodology, and that the goal of mathematics education is a familiarity with logical reasoning, particularly as it concerns quantitative reasoning. We specifically do not include any particular bit of scientific knowledge or any particular mathematical technique. Within the broad domain of science there are far too many items for which a strong case could be made, and we must recognize that only a sample of them could be taught even to the most brilliant children and by an approach depending completely on memorization. Since we hope that most of the effort will be spent on teaching the scientific *method,* we must expect that children will acquire only a small sample of scientific *knowledge.* Moreover, we see no reason why one child's sample should be the same as another's. In considering the extent to which children educate one another, we find a good reason for having several different topics under study within a single classroom. In mathematics there are certainly techniques indispensable to our goal (counting, for instance, or elementary arithmetic), but we feel that emphasis on these as ends in themselves would be misplaced.

Exercises

1 For what demands of 'real life' should we attempt to equip children? Try to find criteria which will help in the identification of such demands.

2 How do the basic philosophies and points of view of mathematics and science differ? Can they truly be correlated or will it be the case that mathematics will be 'integrated out'?

3 Compare and contrast the aims of teaching mathematics, science and creative arts in schools.

4 'Since . . . most of the effort will be spent on teaching the scientific *method* we must expect that children will acquire only a small sample of scientific *knowledge.*' Discuss.

5 Think of some everyday experiences which could prove a starting point for mathematical investigations.

4.   Innovatory strategies

Just as the development of a new concept may build on existing ones, so previously developed innovatory strategies may contribute to the development of new approaches. On the other hand, it may be necessary, before an approach can be implemented, to develop a new model and organisation structure. In the sequence of mathematics projects it is possible to detect a gradual transition from those projects which sought their innovatory model in the political and social environment, to those whose conceptual approach demanded a newly-created, purpose-built model.

4.1.   *The R–D–D model*

Consideration of the various forms of school organisation and administration shows how strongly they influence possible innovatory strategies. In the Federal Republic of Germany the hierarchical and centralised organisation at the level of the Länder has so far, with possibly a few exceptions within the comprehensive school movement, permitted innovation only by decree: innovations imposed at the top of the hierarchical edifice. In contrast, the English and American school systems are characterised by a high degree of local autonomy. The result, particularly in the case of the US system, is to create a vast commercial market at the local level with numerous potential customers. In the circumstances the readiness to look to the world of commerce and industry, and to transfer successful innovatory strategies from it to education was not unnatural. The strategy borrowed from the technological planning and development of new industrial products, known as the 'technological model', divides the process of innovation into the five clear-cut phases of planning, research, development, dissemination and application. The three central phases of research–development–dissemination gave rise to what is frequently referred to as the R–D–D model for innovation.

A feature of this model is that at the end of the development phase there is a product which, tests apart, is not presented to the potential user before the dissemination phase. The too early participation of the user in the process of innovation is, in this strategy, seen as being detrimental to efficiency and quality. These are best served and ensured by the specialised scientists and technicians working on the different phases. Evaluation of the innovation is done in terms of the materials produced by the project, or of their success on the open market.

The R–D–D model and the behaviourist approach to curriculum development proved an ideal match (and not only in the field of mathematics education); learning processes with clearly specified contents and outcomes can readily be conceived in terms of products in the sense of the R–D–D model. Their materials have to be developed and marketed just like industrial products. The behaviourist approach, for its part, found the best opportunity for its realisation in the R–D–D model. No wonder then that the two approaches were successfully paired at an early stage in industrial and military training programmes and that the R–D–D model was adopted by those curriculum developers who attempted further to develop the behaviourist approach out of psychological–pedagogical reflections and isolated applications in school arithmetic, and into self-contained curricula.

The production of taxonomies and their translation into curricula were entrusted to specialists. The teacher had very little part to play; he was concerned in the dissemination phase but application was often the concern of the pupil. Thus in computer-assisted instruction the teacher is there merely to provide help in computer operation.

The behaviourist projects have, therefore, little to offer the teacher in pre- and in-service courses other than instruction in programmed teaching, which assumes great importance because of the strict practice-orientation of the approach. In principle, however, this approach seeks to reduce as far as possible the subjective influence of teachers: the construction of teacher-proof materials is its essential goal.

Like the behaviourist approach, the New-Math approach is also exclusively product-oriented. In this too we can distinguish clearly between three distinct phases: first, professional mathematicians specify the content of the new curricula; second, specialists develop the given content into curricular units in the form of textbooks and courses; finally, these are marketed, usually by commercial publishers.

The translation from concept to practice makes no special pedagogical demands since in this approach new content is seen as the goal of the reform. The project's prime aim as far as in-service education is concerned is to ensure that teachers 'understand the mathematics' and are kept abreast of the latest developments in university mathematics.

The unrestricted use of the R–D–D model of curriculum development was first called into question when it was realised that the teacher had a more vital role in education than he had been assigned in the behaviourist and New–Math approaches. Once the aim shifted to the production of learning situations which were not totally initiated and controlled by outside materials and in which the teacher assumed the unpro-grammable and individual role of an initiator of learning and cognitive processes, the deficiencies of the R–D–D model became apparent. This occurred at a time when, under the influence of research on cognition, discovery learning was attributed greater importance within the struc-turalist concept: a teaching method which, of course, is heavily depen-dent on the teacher's ability to motivate and guide. It was still the case that the 'professional experts' dominated the first phase, for it was they who were responsible for identifying the 'structures of the discipline', a task which it could be claimed had only been successfully accomplished in the one discipline of mathematics. However, in the next phase, the translation of these mathematical prerequisites into curricula, participa-tion of teachers was essential in order to ensure that the teaching models produced were pedagogically feasible within actual classrooms. In-creasingly, 'experts in pedagogy', educationalists and trainers of teachers, were included in project teams. An attempt was made to combine mathematical competency and know-how of school practice; to bring together the mathematical and pedagogical components of educa-tion.

4.2.   *A model for personality-oriented curriculum development*

The formative and integrated-teaching approaches precipitated the formation of an innovatory strategy to replace the R–D–D model. No longer could research, development and dissemination be seen as three distinct phases. In particular, research and development could not remain university-based; the emphasis had to switch to the school and the teacher became the focus of the process of innovation. The new strategy aimed at strengthening the teacher, and at making him better able to function at a professional level and of assuming a creative role within the overall curricular approach. No longer was he to be regarded as the mere performer of a ready-to-use curriculum.

In order to accomplish this it was essential that the teacher should properly understand and adopt the project's aims. This often meant his involvement in the basic, mostly psychological, research of the project. It was equally essential that the materials offered could actually be used in practice, and that they could be successfully handled by the average teacher. The implication was that curriculum development could only be done by teachers, with assistance where necessary from other mathematicians and educators, for teachers. Ideally, the developer and the recipient of the project work should be the same person, so that, as often occurs at university level*, each teacher should develop himself the curriculum that he will have to teach.

The dilemma posed by the translation of such a claim into practice is solved (at least, in theory!)† by supplying materials which are not intended so much for direct classroom use but rather as paradigms. These will then supply the teacher with ideas and starting points. Such curricular units will not be developed in a single place and will have no claim to universality. The development is spread over as many different centres as possible, each centre being able to take into account its own peculiar regional requirements.

In this form of organisation the self-contained working group is replaced by field staff who work as teacher consultants and guarantee the feed-back of experience and suggestions on the project.

The integrated-teaching approach goes yet one step further in seeking to extend participation in curriculum development to *all* concerned, that is, it includes the pupils also.

We see then that in this model of curriculum development there is no longer any separation between 'R and D' and practice: the two are inextricably linked.

In the USA this new model stands in direct contrast to the established forms of curricular reform in which the idea of a link between the developers and users of a curriculum is alien and in which dissemination becomes a commercial marketing process. In England the many teachers' centres established by the local authorities have provided a basis for such curricular initiatives.

In both countries, however, the same dilemma has had to be faced, namely, that projects of this type have still been under an obligation to offer their materials on the open market, thus running the risk that the interaction between curriculum development and practice might simply be confined to an initial service phase.

Exercises

1   Investigate the way in which in the late 1950s and 1960s major US industries bought up publishers and other companies concerned with education. (See, for example, Davis (1967) who names, amongst

---

*   Unfortunately at this level 'curriculum' is still often regarded as synonymous with 'syllabus content'.

†   In practice, disappointments have arisen because the enormity of the task 'strengthening the teacher' and 'making him better able to function at a professional level' has not been granted due weight.

others, CBS, NBC–RCA, General Electric, Raytheon and IBM.) What were the reasons for this? Were expectations justified? What were the consequences for publishers of the move away from the R–D–D model?

2 'There has in recent years been criticism of large, centrally organised projects for their failure to recognise the individuality of schools and for their failure to realise the potential of the teacher as a professional in his own classroom. School-based curriculum development, on the other hand, may suffer through lack of resources and lack of external perspectives and reference points.' (Case study on Ireland; Ruddock & Kelly, 1976.) Exemplify and discuss.

3 Eisner & Vallance (1974) offer five 'conceptions of curriculum: the development of cognitive processes; curriculum as technology; self-actualization, or curriculum as consummatory experience; social reconstruction – relevance; and academic rationalism'. Examine their characterisation of these five approaches. How well do they match the five approaches discussed in this chapter?

4 Davis *et al.* (1978), when considering cognitive processes in learning algebra, use metaphors and language borrowed from the field of artificial intelligence. The work demonstrates a shift in Davis's attitude and does not exemplify a 'formative' approach. How would you describe its orientation? Is such work likely to produce curricular responses that can be classified using the five labels given above, or will it necessitate the formulation of a sixth 'approach'? What other gaps exist in the classification given, or are likely to arise?

6

A Retrospective Look at Curriculum Projects

1.   The reform period in the USA

For many years behaviourist learning theory had little impact on curriculum discussion, for this was dominated by the ideas of Dewey's school whose child-centredness was more appealing than the behaviourist emphasis on the meeting of society's demands. Behaviourist theory was also noticeably deficient in that it had to leave open many questions regarding the hierarchical organisation of the objective goals and their applicability, questions to which answers were not given until Bloom's taxonomy (Bloom *et al.*, 1956) and Gagné's types of learning (1962). World War II was, however, to produce a marked swing in favour of the behaviourist approach. There were several reasons for this.

One was the way in which society's demands on education came to assume far greater significance and to prevail over those objectives which were directed at individual needs. Again, investigations carried out in the armed services into the educational level of their recruits exposed a disastrous ignorance of mathematics and the sciences which was widely attributed to the failure of the liberal–pragmatic education to transmit knowledge and skills. There was an urgent need to repair such failings and here the behaviourist methods of training seemed particularly apposite. Moreover, in the limited way in which they were employed they proved highly effective. Thus it was that behaviourist methods came to be used in military and business circles before they entered schools.

Attempts then followed in a natural and understandable way to extend the scope of their application from local use in isolated sectors to the whole system of education. Moreover, it was not for some years, until after the Korean War, in fact, that the limitations of the behaviourist approach became more and more apparent. Then one of its leading exponents, R. M. Gagné, who had in the 1950s been entrusted with the development of a curriculum for the training of the US Air Force, came to realise that certain training goals could not be achieved by means of the behaviourist approach alone. He was led to suggest some hybrid form of training.

At that time, however, the behaviourist school was, in fact, having to compete with a new reform movement, one with highly specific aims and which drew its inspiration neither from the behaviourists nor from the curricular theories of the progressive era.

1.1. *Projects based on the new-math approach*

The reforms in American pedagogy which occurred in the first half of the twentieth century were theoretically based on philosophy and psychology, and this was also true of the curricular reforms in mathematics and the natural sciences. The content of school mathematics appeared to be built upon a firm and utterly sufficient body of knowledge, largely fashioned by the belief that only those topics that might be of immediate use should be taught in schools devoted to the education of the general public. Thus computational skills and techniques dominated in the primary schools, and, in the secondary schools, simple algebra, trigonometry and such applications as commercial arithmetic. The usual mode of teaching was by the division of these contents into simple tasks, a method which for some time was not seriously to be called into question.

This, in some way, ran counter to Dewey's proposals, but these had, in fact, remained too general for his postulate of 'learning by doing' to spark off significant changes in mathematics teaching. On the other hand, the behaviourists supported this division into tasks and the means of control it offered as a parallel to, and a confirmation of their theory. They sought then to refine this teaching approach, not to question it. In so far as it dealt with concrete contents, the pedagogy of mathematics was concerned with 'task analysis'.

The work of the Bourbaki group, and other developments in the world of mathematics, initially made no impact whatsoever on the mathematics taught in schools. It only resulted in the gap between school and university mathematics becoming even wider. The universities responded to this 'mathematical pressure' for change, but teachers in the schools were largely unaware of what was happening.

However, many attempts in the early 1950s to modernise college introductory courses and to enrich them with more demanding mathematical content were frustrated by the low level of understanding and knowledge of mathematics displayed by the students emerging from high school. It seemed essential that efforts to improve that level had to start in the schools. The mathematics curriculum of the secondary, and later the primary, schools became a matter of concern for the university mathematician.

The result of this concern was a wave of reform in school mathematics which proved to be far more wide-reaching and fundamental than had ever been contemplated in the first instance. For originally the changes envisaged had not been so great, which accounts for the absence of any comprehensive stock-taking or analysis of existing teaching models. The corrections to be made were simply those which manifested themselves to university mathematicians as obvious shortcomings of the system at that time. Questions of method were largely ignored: indeed a previous overemphasis on method was by many held responsible for the neglect of content. 'Content analysis' became contrasted with task analysis.

The beginning of the reform was, therefore, a period marked by great optimism and confidence that the task could be solved quickly and without complicated resources. It was a matter which the mathematicians thought they could handle alone.

The project set up at the University of Illinois by the Committee on School Mathematics, the UICSM, was to prove the prototype for an era of curricular development in the USA. It aimed to improve the teaching of mathematics to pre-college students, for the benefit of the universities, so as to help overcome the gap between school mathematics and that at the university, and to secure a better qualified new generation of mathematicians. After an initial reappearance of the 'old maths' in a 'new dress', that is, re-arranged and newly systematised, more precisely formulated and subjected to an axiomatic development, the New-Math concept gradually took shape. This phase of the UICSM's work served as a model for many other projects; the UICSM was itself to turn to other goals after 1962. With the setting up of the Boston College Mathematics Institute (BCMI) Project in 1957 the New-Math concept was explicitly accepted within the context of teacher education.

The best known, and probably the largest project on the mathematics curriculum to be established in that era was the School Mathematics Study Group (SMSG) which was set up in 1958. The SMSG emerged from two conferences of mathematicians drawn from the major US universities and the procedures which it adopted were typical of those of many of the projects in the initial phase of reform. In particular, it implicitly employed the R–D–D model characteristic of that whole era. Thus, the goals and topics for the courses and the type of materials to be provided for pupils and teachers were quickly determined by the mathematicians. Topics were then prepared and assigned to 'authors', who often produced the texts in the summer vacations. These were then tested in the schools during the next school year, revised and entrusted to a publisher, or published by the developers themselves.

The remarkable 'success' of the SMSG is demonstrated by the fact that it was translated into 15 languages. This success may well have been due to the fact that it was the first project to offer a comprehensive programme in New Math for all levels, from the kindergarten to the 12th grade.

However, if 'success' is measured in terms other than books sold and the number of foreign language editions, then the position is less straightforward. As Vogeli (1976) points out, the project did not meet with the same degree of success in the primary school as it did in secondary education. In the primary area, dissemination problems loom larger. Moreover, the attempts to transfer SMSG materials to other countries created a variety of problems.

Yet the influence of the SMSG was indeed much greater than even the widespread distribution of its texts would indicate. It provided a stimulus and a model to innovators throughout the world (see, e.g. Thwaites, 1961, 1972) and its thinking is to be seen in many publications of the 1960s, for example the 1968 KMK (Conference of Ministers for Cultural Affairs) decisions on the reform of the mathematics curriculum in Federal Germany make explicit reference to it. Moreover, it served to train a generation of new textbook authors, and, as it originally intended to do, exerted considerable influence on subsequent commercial texts. Yet in the diffusion of its work, much was lost. The variety of new work and the rigorously deductive methods were reduced or watered down; and topics such as set theory and algebraic structure lost their role as 'relational' links and became mere inventories of concepts.

Case study

To illustrate the 'New-Math' approach we have chosen excerpts from an early UICSM textbook, Beberman and Vaughan's *High School Mathematics – Course 1.** We notice how little of Beberman's (1958)

---

* Copyright 1964, D. C. Heath and Company. Extracts reproduced by permission of D. C. Heath and Company.

concern for method can be detected in the 'commercial' form of UICSM's materials.*

High School Mathematics

	Introduction Principle	Uniqueness Principle	Definition Principle
Subtraction	$\forall_x \forall_y \ (x - y)$ $+ y = x$	$\forall_x \forall_y \forall_z$ if $z + y = x$ then $z = x - y$	$\forall_x \forall_y \ x - y$ $= x + -y$
Oppositing	$\forall_x x + -x = 0$	$\forall_x \forall_z$ if $x + z = 0$ then $z = -x$	$\forall_x - x = 0 - x$
Division	$\forall_x \forall_{y \neq 0}$ $(x \div y) \cdot y = x$	$\forall_x \forall_{y \neq 0} \forall_z$ if $z \cdot y = x$ then $z = x \div y$	$\forall_x \forall_{y \neq 0} \ x \div y$ $= x \cdot /y$
Reciprocating	$\forall_{x \neq 0} \ x \cdot /x = {}^+1$	$\forall_{x \neq 0} \forall_z$ if $x \cdot z = {}^+1$ then $z = /x$	$\forall_{x \neq 0}$ $/x = {}^+1 \div x$

The above table indicates the systematic, precise and abstract way in which principles were presented.

Traditional applications of mathematics such as percentages and interest were no longer granted an important place within the mathematics curriculum. They received a formal treatment which emphasised only their structural, mathematical aspects and not their use in dealing with problems outside mathematics. Thus, for example:

The symbol '17%' refers to a singulary operation. To find 17% of

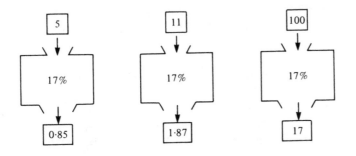

* The reader should note that in this and later case studies the original page and type sizes, and layout have not been preserved.

a number amounts to multiplying the number by 0.17. Thus, for example,

$$17\% \text{ of } 63 = 63 \times 0.17 = 10.71.$$

Also, the number which is 17% greater than 63 is 63 + (17% of 63), and so is 73.71.

Answer each of the following questions.

1. What number is 12% of 73? [$x = 12\%$ of 73]
2. What number is 73% of 12?
3. Find the number which is 20% of 60.
4. Find the number 30% of which is 27. [30% of $x = 27$]
5. What percent of 75 is 15? $\left[x\% \text{ of } 75 = 15 \iff 75 \cdot \dfrac{x}{100} = 15\right]$
6. 37% of what number is 29?
7. What number is 127% of 87?
8. 42 is 157% of what number?
9. 17 is 17% of what number?
10. Find the number which is 10% greater than 30.
11. Find the number which is 28% greater than 41.
12. Find the number which is 15% less than 90.
13. What number is 120% of 50?
14. What number is 120% greater than 50?
15. What number do you get when you increase 40 by 50% [of itself]?
16. What number do you get when you decrease 60 by 50%?
17. By what percent should you decrease 60 to get 40?
18. What is the sum of 50% of 70 and 60% of 30?
19. How many stamps does Phil now have in his Switzerland collection if acquiring 60 new ones would increase the size of his collection by 50%?

Great attention was paid to the presentation of concepts and principles in a precise, clearly defined (and, where necessary, newly-created) language:

Summary of Principles for Real Numbers

Here is a list of the principles for real numbers which we have named so far. For each of these principles, we give an abbreviated name and a pattern for its instances. As you read through the list write at least one instance of each principle.

The Associative Principle for Addition [APA]

$$(\underline{\quad} + \_\,\_\,\_) + \ldots\ldots = \underline{\quad} + (\_\,\_\,\_ + \ldots\ldots)$$

The Associative Principle for Multiplication [APM]

$$(\underline{\quad} \cdot \_\,\_\,\_) \cdot \ldots\ldots = \underline{\quad} \cdot (\_\,\_\,\_ \cdot \ldots\ldots)$$

The Commutative Principle for Addition [CPA]

$$\underline{\quad} + \_\,\_\,\_ = \_\,\_\,\_ + \underline{\quad}$$

The Commutative Principle for Multiplication [CPM]

$$\underline{\quad} \cdot \_\,\_\,\_ = \_\,\_\,\_ \cdot \underline{\quad}$$

The Switch Principle for Addition [SPA]

$$(\underline{\quad} + \_\,\_\,\_) + \ldots\ldots = (\underline{\quad} + \ldots\ldots) + \_\,\_\,\_$$

The Switch Principle for Multiplication [SPM]

$$(\underline{\quad} \cdot \_\,\_\,\_) \cdot \ldots\ldots = (\underline{\quad} \cdot \ldots\ldots) \cdot \_\,\_\,\_$$

The Twist Principle for Addition [TPA]

$$(\underline{\quad} + \_\,\_\,\_) + (\ldots\ldots + \sim\sim) = (\underline{\quad} + \ldots\ldots) + (\_\,\_\,\_ + \sim\sim)$$

The Twist Principle for Multiplication [TPM]

$$(\underline{\quad} \cdot \_\,\_\,\_) \cdot (\ldots\ldots \cdot \sim\sim) = (\underline{\quad} \cdot \ldots\ldots) \cdot (\_\,\_\,\_ \cdot \sim\sim)$$

The Distributive Principle for Multiplication over Addition [DPMA]

$$(\underline{\quad} + \_\,\_\,\_) \cdot \ldots\ldots = (\underline{\quad} \cdot \ldots\ldots) + (\_\,\_\,\_ \cdot \ldots\ldots)$$

The Left Distributive Principle for Multiplication over Addition [LDPMA]

$$\ldots\ldots \cdot (\underline{\quad} + \_\,\_\,\_) = (\ldots\ldots \cdot \underline{\quad}) + (\ldots\ldots \cdot \_\,\_\,\_)$$

The Principle for Adding 0 [PA0]

$$\_\,\_\,\_ + 0 = \_\,\_\,\_$$

The Principle for Multiplying by $^{+}1$ [PM $^{+}1$]

$$\_\,\_\,\_ \cdot {}^{+}1 = \_\,\_\,\_$$

The Introduction Principle for Oppositing [IPO]

$$\_\,\_\,\_ + {}^{-}\_\,\_\,\_ = 0$$

The Principle for Multiplying by 0 [PM0]

$$\_\,\_\,\_ \cdot 0 = 0$$

Vocabulary Summary

commutative	transitivity
associative	distributive
instance	left distributive

consequence	whole number
pattern	whole number factor
derivation	odd number
symmetry	even number
principle	digit

Review Exercises

1. The pair $(^+3, \underline{\quad})$ belongs to the operation adding $^-7$.
2. The pair $(\underline{\quad}, {}^+3)$ belongs to the inverse of adding $^-7$.
3. The inverse of adding $^+21$ is the same operation as adding _____.
4. Subtracting $^+21$ is the same operation as adding _____.
5. $(^+3, {}^+5)$, $(^+6, \underline{\quad})$, and $(\underline{\quad}, {}^-1)$ all belong to the same multiplication operation.
6. Does the operation multiplying by $^-2$ have an inverse?

Write the complete name of each of these principles for real numbers and write the principle.

7. IPS	8. UPS	9. DPDA	10. DPS
11. DPO	12. UPO	13. IPO	14. UPD
15. IPD	16. DPDS	17. DPD	18. UPR
19. DPR	20. IPR		

Which of these generalizations are true?

21. $\forall_x - -x = x$

22. $\forall_x \forall_y - (x - y) = y - x$

23. $\forall_x x^2 = (-x)^2$

24. $\forall_x \forall_y \forall_z x - y - z = x - (y - z$

25. $\forall_{x \neq 0} {}^+2x^2 \div x = x$

26. $\forall_x \forall_{y \neq 0} \dfrac{x}{y} = \dfrac{-x}{-y}$

27. $\forall_x \forall_{y \neq 0} \dfrac{x}{y} = x \cdot \dfrac{^+1}{y}$

28. $\forall_{x \neq 0} \dfrac{x}{x} = {}^+1$

29. $\forall_{x \neq 0} \forall_{y \neq 0} x \div \dfrac{x}{y} = y$

30. $\forall_x \left(\dfrac{x}{^+1}\right)\left(\dfrac{x}{^+1}\right) = \dfrac{x^2}{^+2}$

31. $\forall_{x \neq 0} x \cdot \dfrac{^+1}{x} = {}^+1$

32. $\forall_x \forall_{y \neq 0} \forall_{z \neq 0} \dfrac{x}{y} \div z = \dfrac{xz}{y}$

33. $\forall_x \forall_y \forall_z x - (y + z) = (x - y) - z$

34. $\forall_u \forall_{v \neq 0} \forall_x \forall_{y \neq 0} \dfrac{u}{v} + \dfrac{x}{y} = \dfrac{uy + vx}{vy}$

Exercises

1 To what extent does the UICSM course appear to exemplify the principles laid down by Dieudonné (pp. 101–7)?
2 Write an essay on the place of quantifiers in mathematics and in mathematical education.

Discuss ways in which the notions and symbolism of quantifiers can be introduced.

3   Beberman placed considerable emphasis on precision of language and the avoidance of ambiguity. Discuss the consequences of this (i) for the author, (ii) for the teacher in the classroom.

4   What can be inferred from these extracts about the teaching methods the authors hoped to encourage? Answer the same question about other 'new math' texts such as Papy's *Modern Mathematics*.

5   Take specimen 'New-Math' texts, e.g. UICSM, Papy, early SMSG, and describe the authors' apparent attitudes towards (i) motivation, (ii) applications of mathematics.

6   Investigate how various projects introduced the multiplication of negative numbers. Does their choice of method justify or contradict the type of classification given in this chapter?

7   Discuss the aims of review exercises 7–20 in the case study presented above. What arguments might be advanced for or against the listing of such principles?

1.2.   *Projects based on the behaviourist approach*

University mathematicians had not been the only people to complain of the standards reached in schools. Voices were also raised in the armed forces, commerce and industry. The basis of their complaint, however, was not so much the content, for this to a great extent was fashioned by society's demands, but the failure of the schools to achieve acceptable standards. The inability of school-leavers to apply mathematical rules and their lack of computational techniques was (as ever) cited as evidence of under-achievement on the part of the schools.

Industry saw little prospect of help arriving through the agency of the New-Math projects, for these appeared too élitist to meet its needs; there was a demand for 'marketable skills'. (At this point readers may well note that the progress of mathematical education is not entirely unrelated to that of a continuous film performance!)

The failure of school education was attributed not to the choice of content but to the inefficacy of the teaching methods used; and an entrée was thus provided for educational psychologists, many of whom had honed their skills in the armed forces and industry providing training sequences based on the behaviourist approach. This theory, it was urged, had import for schools; all the more since the development of taxonomies of behavioural objectives and of programmed learning was now proceeding apace. The claims of the champions of behaviourist theory did not seem unconvincing: there, to hand, lay a systematic, self-contained. theoretically-based and well-tested approach which was moreover content-neutral and so usable as the framework for a curriculum however the content priorities were assigned.

In 1957 the UMMaP (University of Maryland Mathematics Project) was established. It sought, under the guidance of Gagné, to develop a total hierarchy of mathematical behaviourist objectives and to translate them into programmed-learning curricula. An attempt was made to relate Bloom's *Taxonomy of Educational Objectives* to mathematical goals and objects in an appropriate manner and to produce a sequence of general and detailed objectives. The project began with the production of materials for the 7th and 8th grade. It soon became apparent that these had to be supplemented with pre- and in-service courses for teachers in which programmed learning was both content and method.

The new content introduced by the New-Math projects was soon assimilated by the behaviourists. Even difficult items of content could be formulated as components of a self-contained system of rules in terms of behaviourist-learning objectives and so made accessible to programmed learning. This apparent conceptual linkage between behaviourist learning theory and the New Math gave rise to a series of projects, for example, the Greater Cleveland Mathematics Program of the Educational Research Council of America (GCMP) and IPI (Individually Prescribed Instruction) Mathematics.

The former project, financed at the outset by small local and regional private foundations, school districts and private businessmen, was so successful in its sales of materials that by 1964 it could itself finance two-fifths of the annual budget of the Educational Research Council which sponsored it at the organisational level. Because of its clear presentation of objectives, its production of teaching sequences and its goal-attainment tests, combined with the appeal exercised by the approach, it became a model for numerous small projects launched in the R and D centres and in the regional educational laboratories.

However, the claim of the behaviourist approach to be universally usable within a total curriculum was later to be abandoned (though not explicitly). Of Gagné's eight hierarchically related types of learning (see pp. 96–7), only the three lowest could be formulated in strict behaviourist terms. No longer were comprehensive behaviourist-based mathematics curricula produced.

However, incentives still existed for projects, centres and laboratories to produce learning units in this mode. In particular, the acquisition of skills and techniques, especially amongst low achievers, was frequently entrusted to the behaviourist method, often in the context of computer-assisted or computer-controlled teaching.

Thus a division of labour between the behaviourist concept and other approaches gradually emerged.

Case study

Possibly the most influential and best documented of the behaviourist projects was the Individually Prescribed Instruction Project of the Learning Research and Development Center, University of Pittsburgh. Various curricula were produced based on sequences of behavioural objectives organised into areas, levels and units. The system works as follows: each pupil is 'placed' in a learning continuum at a point commensurate with his performance level. Thereafter he proceeds at his own pace and is called upon periodically to demonstrate his acquired skills. For each unit of the curriculum there is a criterion-referenced pretest of the objectives included in that unit, thus allowing the child to 'pretest out' of certain parts of the unit. In each unit there are also curriculum embedded tests (CETs) which measure performance on one specific objective in the sequence and act as a summary of that task on which the child has been working. On completing the unit the child takes a 'unit post-test'. Success on the post-test indicates that the pupil has the prerequisites for subsequent units; failure indicates the need for remedial work.

It is the duty of the teacher to administer the various tests and to help prescribe suitable material for the pupils. To help in this task are guides *Diagnosing and prescribing for individualised instruction,* which include detailed case studies (some with accompanying cassettes).

We reprint below:

1 The prescription form for Archie Rhodes (Grade 4) taken from the cover of his 'student booklet' for Skill 1 of E-Fractions (Part 1). The case study explains that Archie's poor showing on CET I was connected with his absence from school for several days. 'The teacher realising he just needs a brief review, talks with Archie and then prescribes from the summary page 13, which Archie completes successfully.' The M in the entry for CET II indicates Archie has achieved 'mastery' of skill 1.

2 The objective and page break-downs for F-Fractions 1. Note that the letters following the page numbers indicate r:review, t:teaching and s:summary.

3 Pages 3–9 of F-Fractions 1.

ipi®

MATHEMATICS

Individually Prescribed Instruction

LEARNING RESEARCH AND DEVELOPMENT CENTER
University of Pittsburgh

RESEARCH FOR BETTER SCHOOLS, INC.
Philadelphia

APPLETON-CENTURY-CROFTS
New York

Name _Archie_

Grade _4_

Prescription Form

Page	Instructional Notes	Number of Points	Number Correct	Pres. Init.	Date
1		6	6	CR	1/9/72
2		8	8		
3		7	7		
4		4	2 ✓		
	TEACHER TUTOR			CR	1/9
6		13	13		
7		12	9		
	TEACHER MADE SKILL SHEETS			CR	1/10
8		4	4	CR	1/10
9	CET I	6	0		1/18
13		4	4	CR	1/18
14	CET II	9	9 M	CR	1/18
	good!				
	go get skill 2				

INSTRUCTIONAL TECHNIQUES

Setting	Materials
Peer Tutor	Curriculum Tests
Small Group	Teacher-made Skillsheets
Independent Study	Records Tapes Film Strips
	Manipulative Devices

ipi®
MATHEMATICS

F–Fractions–1

OBJECTIVE: Given two common fractions less than or equal to 1, the student renames each fraction using the least common denominator for the given pair and writes >, <, or = between the given fractions to show their relationship.
LIMIT: Given fractions have denominators ≤ 50.

PAGE	DESCRIPTION
1r	Compares fractions with like denominators.
2t	Renames fractions with a given number as denominator.
3t	Renames fractions with a given number as denominator.
4r	Finds the L.C.M. using factorizations.
5r	Finds the least common multiple of two numbers. Introduction to the term least common denominator
6t	Compares fractions with and without like denominators.
7t	Compares fractions with unlike denominators.
8s	Compares fractions with unlike denominators.
9	CET I
10r	Finds the L.C.M. using factorizations.
11t	Renames fractions with a given number as denominator.
12t	Compares fractions with unlike denominators.
13s	Compares fractions with unlike denominators.
14	CET II

Order Number:
F-08-1

F – FRACTIONS – 1

F – FRACTIONS – 1

3 | no. points | 9 | | no. correct |

A number multiplied by one is equal to itself.

Rename $\frac{2}{5}$ as a fraction with denominator 15.

$$\frac{2}{5} = \frac{?}{15}$$

Since 5 X ___3___ = 15, use $\frac{3}{3}$ as the fractional name for 1.

$$\frac{2}{5} \times \frac{3}{3} = \frac{6}{15}$$

Rename $\frac{2}{5}$ as a fraction with denominator 25.

Since 5 X _____ = 25, use $\frac{5}{5}$ as the fractional name for 1.

$$\frac{2}{5} \times \frac{5}{5} = _____$$

Rename $\frac{2}{5}$ as a fraction with denominator 100.

Since 5 X _____ = 100, use $\frac{20}{20}$ as the fractional name for 1.

$$\frac{2}{5} \times _____ = _____$$

| no. points | 7 | | no. correct | 4 |

Fill in the blanks.

L.C.M.

Multiples of 6	0	6	12	18	24	30.
Multiples of 10	0	10	20	30	40	50

The chart above shows that the least common multiple of 6 and 10 is _____ .

> Do you remember another way to find the least common multiple?
>
> You can build the L.C.M. with the prime factors of the given numbers.

First write the prime factorization of each composite number.

$$6 = 2 \times 3 \qquad\qquad 10 =$$

Since the L.C.M. must be divisible by 6 and 10, the factorization of the L.C.M. must contain the factorization of 6 and 10. But we do not use any prime more times than it appears in either factorization. Write the factorization for the greater number in the rectangle.

Does (2 × 5) contain the factorization of 6? _____
(Yes, No)

Then write the prime factor of 6 that is not already in the rectangle.

The L.C.M. is [2 × 5] or _____ .

F – FRACTIONS – 1

5 | no. points | 8 | | no. correct |

Later in this booklet you will be rewriting pairs of fractions using their Least Common Denominator.

The least common denominator is the least common multiple of the denominators.

> The least common multiple of a set of numbers is the least nonzero multiple of the numbers.

Since the L.C.M. is a **multiple** of both numbers, both numbers must

d _____ the L.C.M.

You can build the L.C.M. from the prime factors of the two numbers.

Find the least common multiple of 9 and 12.

1. Write the prime factorization of each of the numbers that is composite.

   $9 = 3 \times 3$ $12 =$

2. Since the L.C.M. is divisible by 9 and 12, the factorization of the L.C.M. must contain factorization of 9 and 12. Write the factorization for the **greater number** in the rectangle.

   | |

3. Does (2 X 2 X 3) contain a factorization for 9? _____

   Yes, No

4. If 9 is to divide the L.C.M., what factor must be included in the rectangle

   above? _____

5. The least common multiple is | 2 X 2 X 3 X 3 | or _____ .

F – FRACTIONS – 1

| no. points | 9 | | no. correct | | 6 |

You can compare two fractions with the same denominator. Write $>$, $<$, or $=$ in the box to show the relationship.

$\dfrac{2}{5}$ ☐ $\dfrac{3}{5}$ $\dfrac{4}{5}$ ☐ $\dfrac{4}{5}$ $\dfrac{2}{9}$ ☐ $\dfrac{1}{9}$

Complete the sentence below with the words lesser or greater.

When comparing two fractions with the same denominator, the fraction with the ___greater___ numerator is the _____ fraction.

Fill in the blanks and then write $>$, $<$, or $=$ in the box.

If you want to compare fractions with unlike denominators, it is often helpful to rewrite both fractions using their least common denominator (L.C.D.) as the denominator.

Suppose we want to compare $\dfrac{5}{7}$ with $\dfrac{4}{5}$.

The least common denominator is _____ .

$\dfrac{5}{7} = \dfrac{\bigcirc}{35}$ $\dfrac{4}{5} = \dfrac{\triangle}{35}$

$\dfrac{5}{7}$ $<$ $\dfrac{4}{5}$

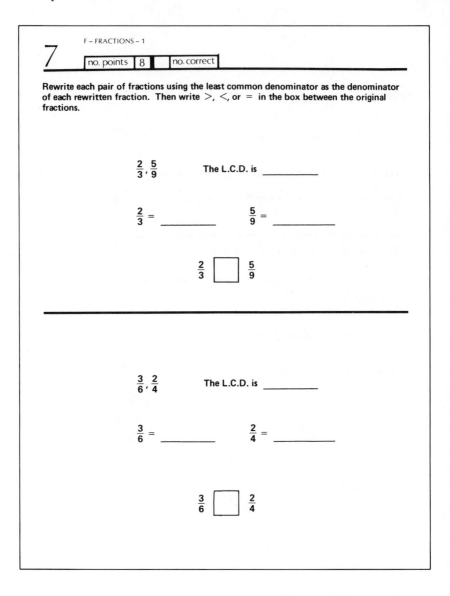

7   F – FRACTIONS – 1

no. points [8]   no. correct

Rewrite each pair of fractions using the least common denominator as the denominator of each rewritten fraction. Then write >, <, or = in the box between the original fractions.

$\frac{2}{3}, \frac{5}{9}$ The L.C.D. is _____

$\frac{2}{3} =$ _____ $\frac{5}{9} =$ _____

$\frac{2}{3}$ ☐ $\frac{5}{9}$

$\frac{3}{6}, \frac{2}{4}$ The L.C.D. is _____

$\frac{3}{6} =$ _____ $\frac{2}{4} =$ _____

$\frac{3}{6}$ ☐ $\frac{2}{4}$

F – FRACTIONS – 1

| no. points | 6 | | no. correct | | 8 |

For each pair of fractions you are to do <u>two</u> things.

1. Rewrite both fractions using the least common denominator as the denominator.

2. Write $>$, $<$, or $=$ in the box between the given fractions to show the relationship.

$\frac{5}{10}$ ☐ $\frac{3}{4}$ $\qquad\qquad$ $\frac{5}{7}$ ☐ $\frac{2}{3}$

$\overline{}$ $\overline{}$
20 $\qquad\qquad$ 20

$\frac{3}{8}$ ☐ $\frac{8}{12}$ $\qquad\qquad$ $\frac{4}{13}$ ☐ $\frac{3}{5}$

$\frac{9}{12}$ ☐ $\frac{3}{4}$ $\qquad\qquad$ $\frac{2}{6}$ ☐ $\frac{11}{33}$

9 F – FRACTIONS – 1 CET I

no. points | 6 | 5 | 4 | 3 | 2 | 1 | 0 |

For each pair of fractions you are to do <u>two</u> things.

1. Rewrite both fractions using the least common denominator (L.C.D.) as the denominator.

2. Write $>$, $<$, or $=$ in the box between the given fractions to show the relationship.

$\frac{3}{4}$ ☐ $\frac{4}{7}$ $\frac{7}{8}$ ☐ $\frac{5}{6}$ $\frac{3}{10}$ ☐ $\frac{2}{6}$

$\frac{5}{9}$ ☐ $\frac{7}{12}$ $\frac{3}{4}$ ☐ $\frac{6}{8}$ $\frac{4}{7}$ ☐ $\frac{5}{8}$

STOP

Exercises

1 It can be argued that one of the aims of any curricular programme should be to foster the mathematical and pedagogical growth of the teacher. To what extent is this a valid argument? In what ways would the IPI procedures encourage the teacher's growth?

2 What can be inferred from the extract about the teaching methods it is hoped to encourage?

3 What would appear to be the project's attitude towards (a) motivation, and (b) applications of mathematics? (Investigate this further by reading other behaviourist material.)

4 Take a chapter from a non-behaviourist text and attempt to provide 'objectives', 'limits' and 'descriptions' of the chapter's contents as it unfolds, in the way which is done for F-Fractions 1.

5 Obtain details of the IMU system (an English version has been published by Caffrey, Smith) and investigate the similarities and differences between it and IPI Mathematics. In particular, consider the advantages and disadvantages of offering alternative teaching tracks.

6 Discuss the language problems inherent in the IPI material.

1.3.   *Projects based on the structuralist approach*

The premises of the behaviourist learning theories – that learning and learning processes can be comprehensively described in terms of objectively observable changes in behaviour, and that it is possible to list all learning processes pursued in schools by means of a general catalogue of learning goals – have been attacked since their inception. The behaviourists have been accused of reducing learning to the transmission of ready-to-use knowledge and techniques and of making it purely passive. In particular, they appeared to ignore the creative aspects of the learning process which were stressed by the 'Gestalt' psychologists. Further opposition to the behaviourist position arose out of the investigations of cognitive-oriented learning theorists into concept formation.

The behaviourists had been able to make the transition from traditional to new content, but in doing so they had failed to convince more than a few university mathematicians that theirs was a suitable method. Indeed doubts and worries were reinforced by the way in which the first behaviourist projects had succeeded only in producing trivial operationalisations, inadequate to mathematical structures. Moreover, even if the behaviourist concept was in keeping with the exactness of the New Math it ran counter to its general spirit and aims. It was not the kind of approach likely to produce a new generation of creative researchers or even questioning citizens. Hence, many who at the beginning of the reform era were primarily interested in the content of the mathematics curriculum began to consider more earnestly the methodological problems of content transmission. How should fundamental scientific concepts be transmitted to the pupil, and what was the relation between the 'elementary' in a scientific sense and that in a school context?

Two approaches came close to an answer.

The first attempted to gain insights into the transmissibility of content by means of an analysis of its structures. The other concentrated on the central issue of cognitive research; how concepts and thinking abilities are acquired and how cognitive processes can best be arranged to enhance learning. The two approaches were synthesised in the early sixties in Bruner's 'structures of the disciplines'.

For the first time a curriculum theory was developed which could claim to deal in a unified manner with problems of content and method, and which laid stress on their interdependency.

The projects based on the structuralist concept are characterised, so far as content is concerned, by an astonishing choice of topics. This was partly a consequence of the highly imaginative teaching models generated and which were thought likely to promote learning by discovery. It also arose from the place occupied by content in the sequential arrangement of the curriculum. For, in accordance with the logic of the approach, topics usually considered 'difficult', because of their degree of abstraction, for example, groups and linear spaces, could now be placed near the beginning of elementary education. The principle on which the construction of materials was based was that they should involve mathematical structures which were then to be discovered by the pupils by abstraction from the various embodiments, such as stories, games and structured materials. So far as the developer was concerned the structures were his starting point and it was his task to construct suitable embodiments; the goal of the child's learning by discovery being to derive from these their abstract content.

It was characteristic of structuralist projects that they were often tailored to the demands of low achievers or handicapped pupils. Initially, this may well have been caused by a desire to illustrate the motivational potential of the approach.

A typical example of the structuralist project is the UICSM in its second phase (since 1962).

The second phase of UICSM exercised a world-wide influence; in particular, it induced research work in Federal Germany on the development of teaching models for the elementary and secondary levels, and it also affected the project work centred upon Z. P. Dienes which took place in Australia, New Guinea, Canada and elsewhere. In the USA, the project helped to bring about the 1963 meeting held in Cambridge, Massachusetts, which collected together many of those involved in curricular reform in mathematics in the USA and sought to arrive at a consensus on the orientation and programme of a long-term reform. For the first time in the history of the new reform movement, it was possible to gain general acceptance* for proposals directed towards a renewal of the whole mathematics curriculum. The outcome of the conference, *Goals for School Mathematics* (see p. 109), was to serve as a guideline for future projects.

The 'goals' envisaged in the publication related to:

1. The general goals of modern mathematics teaching.
The discussion deals, *inter alia*, with 'skills versus concepts', 'technical vocabulary and symbolism', with the relation between pure and applied mathematics in curriculum development and with the potential and limitations of mathematics.

2. Pedagogic principles and techniques.
The 'discovery approach' is presented as the central teaching method; the role of exercises and the need for open-ended problems and for free discussion among pupils is discussed; all being intended as a contrast to drill.

3. The content of mathematics teaching
Arranged according to the different grades, the catalogue of topics contains many demanding topics, a kind of mini-version of university mathematics, which at secondary level has a strong New-Math bias.

The proceedings of the 1963 Cambridge Conference are essentially based on the structuralist approach and in them a position counter to that of the behaviourists is adopted. Learning by discovery had come to be recognised as a method appropriate to mathematics teaching as viewed from a structuralist basis. A balance seemed to be achieved between the choice of content (by professional mathematicians) and the methods advocated, which were apparently based on a theory of cognition.

This unity was, however, quickly to be shattered as concern for cognitive theories and their methodological applications gathered a

* General acceptance did not extend to non-participants. See, for example, Stone (1965).

momentum of its own. The rediscovery – or, even, discovery – of Piaget's research, and especially of his studies of the formation of mathematical concepts and of their generalisation into his theory of operative intelligence, was responsible for opening up a new field for investigation by those projects concerned with the development of cognitive processes and their enhancement through teaching methods. They provided a new focus of interest since, for the moment, it was (naïvely) assumed that the content problem of curriculum development had been solved by the new structuring of mathematics.

Case study

The graded approach of the structuralists, based upon a 'spiral' curriculum, makes the choice of a case study rather difficult: a brief extract cannot show the 'grand structure'. Nevertheless, the following excerpt from the CSMP materials illustrates several key features. It is the material intended to be used in the twenty-eighth lesson of the second semester in the first grade. (The Eli story is then picked up again in the thirty-fourth and forty-seventh lessons.) We note the use of a simple model to introduce the idea of 'negative integers', although this technical term is never employed. The very early, informal introduction of such a key structure as the integers is, of course, in keeping with a structuralist approach.

S28.1   Eli's Magic Peanuts #1

CAPSULE LESSON SUMMARY

Introduce negative numbers via a story about Eli the Elephant, who is confused by magic peanuts he finds. (When a magic peanut meets a regular peanut, both disappear.) Explore the situation further in number stories such as "$5 + \hat{5} = 0$" and "$7 + \hat{4} = 3$", where the notation "$\wedge$" is used to represent magic peanuts.

MATERIALS

Teacher: None
Student: None

DESCRIPTION OF LESSON

T: There is an elephant named "Eli" who lives in the jungle and is always very hungry. What do you think is Eli's favorite food?

Accept suggestions from the students.

T: Eli's favorite food is peanuts. He likes peanuts so much that he always carries a little bag of peanuts with him wherever he goes. One day, while walking through the jungle, Eli spotted a special peanut bush he had never seen before. Eli didn't know it, but the peanuts from this bush were magic!

Eli gathered some of the magic peanuts and put them in his bag with the other peanuts. What do you suppose is so special about magic peanuts?

Allow the students to discuss this briefly.

T: Whenever a magic peanut comes near a regular peanut, both peanuts suddenly disappear!

Draw this picture on the board.

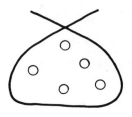

T: This is Eli's bag with five regular peanuts in it. But Eli also put
        some magic peanuts into his bag; here are the magic peanuts.

Continue your drawing.

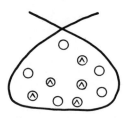

T: How many magic peanuts did Eli put into his bag?

S: Five.

You may wish the class, as a group, to count each magic peanut as you point
to it.

T: When Eli returned home, he was very hungry from walking through
        the jungle all day; he decided to eat some peanuts. When he
        opened his bag, he was very surprised. How many peanuts were
        in the bag, do you think?

S: None; the bag was empty.

S: All the peanuts disappeared!

T: That's right. There was one magic peanut for every regular peanut,
        so all of the peanuts disappeared.

Add these connecting lines to your picture, pairing magic peanuts with regular
ones. Next to the picture, write the appropriate number story.

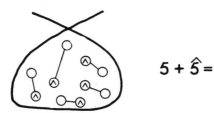

$$5 + \hat{5} =$$

T: Five regular peanuts plus five magic peanuts is . . . ?

S: No peanuts.

S: 0!

Complete your number story on the board.

$$5 + \hat{5} = 0$$

T: Poor Eli was very puzzled. He didn't know the secret of the magic
            peanuts and he couldn't imagine where his peanuts had gone! He
            was still very hungry, so he went looking for more peanuts. This
            time, he found seven regular peanuts and put them into his bag.

Erase your previous picture and draw this new one.

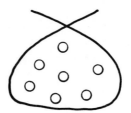

T: Without knowing it, Eli also put four magic peanuts into his bag.

Add these four magic peanuts to your drawing.

T: When Eli returned home, what do you suppose he found when he
            opened his bag?

Ask several students to explain their answers, and then choose volunteers to
pair up regular peanuts with the magic ones.

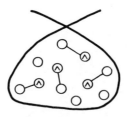

Write this number story next to the picture.

$$7 + \hat{4} =$$

T: Seven regular peanuts plus four magic peanuts is . . . ?

S: Three!

T: Three what?

S: Three regular peanuts.

Complete the number story.

$$7 + \hat{4} = 3$$

Exercises

1   The Eli model for directed numbers allows us to interpret addition but fails to help so far as multiplication is concerned.
    Provide examples of other models which suddenly 'break down'.
    What are the consequences for the use of such models in our teaching?
2   How would you choose to develop the Eli story when next it was encountered on the 'spiral'?
3   What arguments can you see for and against the introduction of directed number so early in the school curriculum?
4   Find other models used for the introduction of directed number and contrast them with Eli (bearing in mind the suggested 'age' of introduction).
5   The Eli model mirrors the definition of an integer as an equivalence class of ordered pairs of naturals. How essential is it that a teacher using the Eli model should have met and understood this abstract definition?
6   The material is very prescriptive so far as the teacher is concerned. Discuss the need for, and consequences of, this.

1.4.   *Projects based on the formative conception*

Developmental psychology, which in the structuralist approach had only played a subsidiary role, now, because of the new insights gained by researchers, assumed a more central role in curriculum research. The recognition of the need to involve teachers more closely in development work and the subsequent emphasis on classroom practice also reinforced the interest in teaching methods.

Moreover, emphasis on content gradually changed. The need to operate within the classroom in accordance with Piaget's theory led to the reinstatement of certain classical types of applications of mathematics, and to a reappraisal of the science-orientation of content. The relative positions of scientific disciplines and cognitive theories in general changed, and gradually the *formative conception* began to emerge and separate itself from the structuralist approach. This transformation manifested itself typically within the long-established Madison Project.

The proposals of this project concerning open curricula, the versatile and imaginative materials produced to encourage its view of the teaching process, and its stance against the translation into schools of university content, with a corresponding loss in that content's substance, were all admired. Yet the approach was to have a greater impact on curriculum theory than on school practice.

Other projects which, sometimes under the influence of the Madison Project, sought similar goals were the Nuffield Mathematics Project in England and the Elementary School Science Project in the USA. The project's views also permeated the reports of two further meetings organised by the Cambridge Conference on School Mathematics: *Goals for Mathematical Education of Elementary School Teachers* (1967) and *Goals for the Correlation of Elementary Science and Mathematics* (1969).

Yet the attempts by the project to bring about changes in actual classroom practice were largely unfulfilled: the discrepancy between on the one hand the goals, and on the other conditions and perceptions as they existed, was too great to be easily bridged. The demands made on teachers were great, and the very nature of the approach was such that teachers could not be given ready-made support in the form of pupils' materials. Moreover, the notion of 'the teacher as a developer' was at odds with that of the 'teacher as a transmitter': a view still held by many teachers.

Case study

The Madison Project was established by university mathematicians engaged in teacher training, who initiated classroom experiments to explore how interest and mathematical activity, creativity, and discovery were best fostered. A consequence of their findings was that the teacher had to be assigned a central role in the process of instruction, and accordingly in curriculum development. Ready-to-use materials would not bring about the desired changes: efforts had to be aimed rather at improving the professional competence of the teacher. For this purpose the project produced films of teacher-discussions and classroom work. A variety of materials intended to assist the teacher in stimulating classroom activity was also produced. Again, the aims of the Madison Project required a new disseminatory strategy based on direct contact with teachers. Accordingly, a network of seminars, consultants and courses resulted.

We illustrate the work and ideals of the Madison Project by reprinting excerpts from notes on pedagogy and the curriculum written by R. B. Davis for the teachers' commentary to the project's films.

Some Questions of Pedagogy and Curriculum

In this chapter we look briefly at seventeen matters of pedagogy and curriculum that are especially important in helping young children learn mathematics. These suggestions come from three sources: the practical experience of classroom teachers; the developmental psychology of Jean Piaget and his colleagues; and a notion of what is important in mathematics, derived from the experience of professional mathematicians, and from scientists, engineers, designers, and others who make extensive use of mathematics.

1. *A developmental approach.* A child's view of the world changes as he gains experience. For younger children, then, the adult version of an idea will often be inappropriate. In place of the usual adult version we need something else, something appropriate to the developmental stage of the child. What can this be? One excellent answer is provided by the 'intermediate inventions' developed by Page, Sawyer, and others.

As suggested [earlier], perhaps the single most distinctive feature of the lessons presented here is their deep commitment to *a developmental approach.* This is not universally accepted as an educational principle, but we would argue that – especially with younger children – it is absolutely essential for sound education.

For clarity, let us state the opposite position, although we reject it.

The opposite, or non-developmental, approach sets as a goal the task of 'getting everything right the first time around' – that is to say, a child is introduced to *adult* versions of an idea if he is introduced to the idea at all. Our own position – the developmental position – holds that presenting young children with adult versions of ideas is in fact impossible. . . .

2. *Paradigm lessons.* The word 'paradigm' has many meanings; we use it here in the sense that a 'paradigm lesson' presents a child with a clear, concrete *example* of something that we want him to think about. A 'paradigm', then, is a clear example of an idea, often presented in dramatic form. . . Pasting ten beans on each of several bean-sticks is a paradigm for 'grouping ten ones and treating them as one ten'. Rather than *telling* children about a mathematical concept, we usually prefer to arrange for them to *experience* the concept in the form of some dramatic paradigm.

3. *Avoiding the perils of ordinary English.* The phrase 'grouping ten ones and treating them as one ten' is a good example of the difficulty of *talking* about mathematics *when one is trying to convey a new idea.* Ordinary English usually fails as a way of telling people about a new mathematical idea, just as it usually fails when one attempts to describe colors to a blind person or sounds to someone who cannot hear. Yet much educational effort is devoted to this largely futile endeavor. Children learn new mathematical concepts by *experiencing* them and by *thinking about what they are doing* – not by being told.

4. *Fighting clear of the ineffective trio.* Three aspects of tradition (and often 'new math') curricula fit together into an inter-locking relationship that stands in the way of significant improvement. One, mentioned above, is trying to *tell* children new mathematical ideas; a second is the inappropriate sequencing of topics in the curriculum – for example, introducing problems such as $1/2 \div 3/5$ before children are ready to *understand* what they are doing; a third is rote learning, where children learn to follow a definite sequence of steps without understanding *why* each step is performed when it is. Obviously, each of these practices makes the others more likely; and reinforcing one another as they unfortunately do, the three are often able to defeat progress. . . .

A major goal for the improvement of school mathematics is to avoid all three of these practices. Instead, one needs to work out an appropriate sequence of mathematical ideas, using the criterion of requiring *understanding*, not memorization; and then to devise learning experiences . . . that ensure that children do develop appropriate understandings of these concepts.

5. *Precise definitions are conspicuous by their absence.* From the

preceding discussion, it is probably clear why the Madison Project lessons usually omit all precise verbal definitions of new mathematical ideas, replacing them by *experiences* that dramatize the ideas and by subsequent discussions of these experiences, cast in a form suitable to the child's developmental state – 'how far he has progressed in his thinking, and what steps he may be ready for next'. Because this order of events is so common in Project lessons, we refer to it as the 'do, then discuss' sequence for developing new ideas. . . .

This is very different from the practice of 'beginning with precise definitions'. What, after all, is a *definition*? In its fully-developed, precise form, a definition represents a sharply-delineated description of exactly which things we shall call *functions* (or rational numbers, or whatever), and exactly which things we shall not call *functions* (or whatever). Precise definitions are used as a way of organizing adult mathematical theory in order to make it possible to construct careful proofs and to arrive at reliable abstract generalizations. A child's knowledge is not organized in precisely this way, and he usually does not need the special precision of mathematical definitions.

6. *The 'meta' principle.* A partial converse of the preceding principle also applies. If we have said that we do not usually discuss things with children until after they have experienced them, the converse is also true: *anything that we have previously experienced* is legitimate grist for the discussion mill. This means that after children have had experiences in making up complicated mathematical statements by elaborating on simple ones, we can subsequently discuss how this is done. . .

Thus, our concern for doing what is appropriate for young children at their present state of development *does not preclude theoretical discussions,* provided that we are talking about things that the children have already experienced.

7. *Belief in intuitive knowledge.* Most of these lessons assume that ideas can develop by moving through a pre-conscious or 'intuitive' stage, then gradually becoming more explicit. This is not the only mechanism by which new ideas can develop, but it can be an important one. Consequently, we take pains both to provide enough experience so that children can develop intuitive ideas; and also to allow them opportunity to tell us about these ideas, describe them, clarify them, and thus render them increasingly explicit.

8. *Mathematics as 'A Story About Reality'.* If the 'developmental curriculum' is the most prominent distinguishing feature of these lessons, a second, almost equally prominent, feature is a commitment to the idea that *mathematical statements describe reality; therefore, mathematics consists of 'stories about reality'.*

(We are speaking about young children in ordinary school settings. We believe that this view is preferable to seeing mathematics as meaningless manipulations of incomprehensible scribblings on paper – which is how far too many children do view it. We do not mean to deny Marshall Stone's remark that at later stages mathematics 'frees us from the constraints of reality' by allowing us to develop so careful a rhetoric that we can safely describe worlds that do not exist, but which would necessarily *have* to be exactly as we describe them *if they did exist.*) . . .

9. *Decisions made by the children.* Whenever convenient, *we allow children to make decisions,* rather than having adults make the decisions for the children. Thus, if we are going to do an addition problem, we typically ask the children to decide what numbers we will add. . . .

10. *The 'tool' concept.* Every mathematical idea is, in effect, a *tool* that was developed for some definite purpose. We want children to experience mathematics in this historically correct way. Consequently, we want mathematical ideas to grow naturally out of problem situations. An important goal of good mathematics instruction is to have the child develop a deep conviction, based on his own experience, that *mathematics consists of reasonable responses to reasonable questions.* This means, in general, that any topic which might seem arbitrary had probably better be deferred until a point at which it can be experienced as a 'reasonable response to a reasonable question'.

11. *Clarifying the 'meaning of the question' vs. 'solving the problem and having the child imitate you'.* When we want a child to learn how to do something mathematical that is unfamiliar to him, we often have two choices: we can make the task very clear and leave the child to solve it by methods he devises himself; or we can ignore the meaning of the task and solve the problem ourselves, asking the student to imitate our method. Both approaches are defensible, but in a good math program we would expect the first method to predominate. . .

12. *Careful observation of cues from the children.* Which is more important: the signals that go *from* the teacher *to* the children, or the signals that go *from* the children *to* the teacher? The *from*-teacher-*to*-child signals include what the teacher *says,* his gestures, posture, intonations, etc. Conversely, the *from*-child-*to*-teacher signals include what children say, what they write, their hesitations, facial expressions, posture, meaningful glances, and so on. Conventional wisdom says that the from-teacher-to-child signals are the important ones and focuses attention mainly on them. We would argue that this imbalance needs to be shifted towards far more concern with the signals *from* the child

to the teacher. A major part of a teacher's job is diagnostic. . . We would argue that a teacher is not a lecturer, nor an actor, nor a town crier. He is, among other things, a diagnostician, a helper, and a model of appropriate adult behavior.

Which would bother you more: a slip of the tongue, where you (as teacher) say something other than what you meant, or a situation in which you misread the meaning of a student's gesture, glance, or remark?

In any event, the distinction between those signals that go from the teacher to the children, *vs.* those that go the other way, is a point worth keeping in mind as you watch the filmed lessons. . . .

Exercises

1 Exemplify further the points advanced by Davis.
2 Write an extended essay on any one of the twelve points quoted above.
3 What are the 'intermediate inventions' to which Davis refers in point 1? How do these differ from Dienes' embodiments? Is there a difference between these latter and Davis's 'paradigms' (point 2)?
4 Write an essay on 'the perils of ordinary English'. (Do not feel constrained by the very specific arguments given by Davis under this heading.)
5 'Definitions are the working hypothesis of the child; they develop gradually with the growth of knowledge.' (Branford, 1908.) Discuss, compare and contrast Branford's views with those of Davis. (Those who have read Branford's pioneering book might care to consider which of the approaches described in this chapter would have most attracted him.)
6 Describe the particular problems of dissemination arising in the case of projects linked with the 'formative approach' and suggest methods by which they might be eased.

1.5. *Projects based on the integrated-teaching approach*

In principle, the more a project deviates from existing conditions the less likely it is to succeed in having its ideas implemented. The only hope lies in realising this from the outset and in designing and adopting an innovatory strategy which recognises and explicitly seeks to alleviate the resulting tensions. In this respect, the integrated-teaching and the formative approaches face similar problems.

As we have seen (chapter 5), the integrated-teaching approach takes as its starting point the general needs and interests of the pupils and relates its choice of content and methods to this.

The major US project based on this concept is the Unified Science and Mathematics for Elementary Schools Project (USMES). This was established in 1970 at the Newton (Ma) Education Development Center, following the recommendations made by the authors of the *Goals for the Correlation of Elementary Science and Mathematics.*

USMES developed further the ideas of the Madison Project, adding to its conception the consideration of content in terms of real problem solving. In order to facilitate the innovation process, USMES, however, relied more on the provision of concrete materials for both teacher and student. Its products are addressed to elementary and junior high schools.

A most significant difference from Madison is that the products are organised by themes. USMES offers 'material-packages', of which 26 were developed. Every such unit refers to a field of concrete problems taken from the world around, e.g. 'Pedestrian Crossings', 'Describing People', 'Classroom Management', 'Consumer Research', 'School Zoo', 'Nature Trails', 'Weather Predictions'.

The units can be completed by teachers and pupils. There is no hierarchic order, e.g. they are not arranged according to degree of difficulty. Most of the units can be used in any grade between 1 and 8. In fact, their function is to challenge the student's interest and to stimulate investigations in which various aspects of the sciences are integrated. Such work should be carried out at the highest possible level consonant with the student's cognitive level. In this way the student will provide real solutions to real problems.

Both teacher and student are supplied with a wide range of materials. The teacher is given a 'resource-book' for every unit, providing general and specific information and examples of how the unit has been carried out by others. Moreover, there are 'background papers' from which the teacher may obtain technical information on a higher scientific level, and a corresponding 'Design Lab Manual', advising the teacher in crafts and practical techniques. A 'Curriculum Correlation Guide' facilitates integration of USMES work with other existing curriculum products.

When working on these units, the students will require many types of help, most of which will be provided by the teacher. It is expected that during the investigations subsidiary problems will arise and the need for new skills will become apparent. This need is partially met by the provision of special 'how-to' cards.

Case study

To demonstrate materials of the integrated-teaching approach is an even more difficult task than it was for the formative approach, because of their variety and richness. Indeed, it appears to be the case that the richer and more colourful materials are, the harder it is to provide representative case studies. Nevertheless, we try to give an idea of the classroom activities which can be stimulated by some short descriptions of units adapted from the guide *Mathematics and the Natural, Social and Communications Sciences in Real Problem Solving (1976)*, published by the Unified Sciences and Mathematics for Elementary Schools Project of Education Development Center.

Eating in school

Challenge:
Promote changes that will make eating in school more enjoyable.

Possible Class Challenges:
How can we improve the lunchroom environment?
How can we improve the service in the lunchroom?
How can we improve the food we have for lunch (snack)?

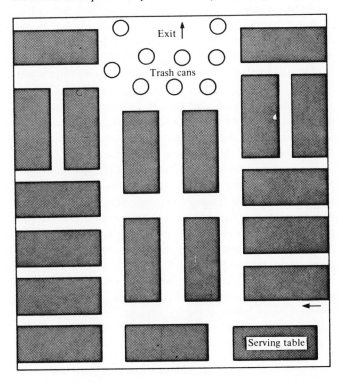

Lunch and breakfast are often sources of student criticisms. Complaints usually focus on the lunchroom itself (too noisy, poor decor, poor furniture arrangement) or the food (unappetizing, poorly served). Whatever the problem(s), students will respond eagerly to the unit challenge because eating is one of the more favored school activities.

Problems connected with their lunch, breakfast, or snack may arise naturally when students complain that the food was terrible or the wait in line was unusually long. In some classes the teacher need merely ask, 'How was lunch or breakfast today?' for a lively discussion to evolve.

While discussing the challenge, the class may list possible problem areas, which may include physical aspects of the room itself, such as mealtime scheduling, the lunch and breakfast lines, traffic flow in the room, furniture arrangement, rules, lunchroom decor, noise, lighting and temperature, and student manners. Other problems may deal with nutritional aspects of food, such as quality, quantity, appearance, manner of serving, student preferences, waste, and storage of cold lunches. To assess the present situation, the students may first observe in the lunchroom for a period of time. The class may then identify one or two urgent problems and choose to tackle them before the others. To prove that the problem exists, they may decide to collect data, e.g., the amount of time it takes a student to get through the line, the number of students eating lunch or breakfast during each time period, the kinds of foods students throw out, the way the food is laid out and served.

For some of the tasks, the class may divide into small groups. As the lunchroom is used by the whole school, the students may conduct opinion surveys to determine preferences and additional criticisms. Any data collected through surveys and observations are graphed in preparation for a presentation to the school faculty, staff, and principal; scale layouts of the lunchroom may also be made. Many revisions of proposed changes may take place before a trial implementation of the plans occur. During the trial implementation period the students collect data in order to assess the effectiveness of their plans. Revisions and surveys are once again made before the final plans are put into effect.

The students' interest may extend to improving other areas of the school, and they may then decide to pursue such USMES challenges as School Rules, Classroom Design, Classroom Management, or Play Area Design and Use.

Pedestrian crossings

Recommend and try to have a change made that will improve the safety and convenience of a pedestrian crossing near the school.

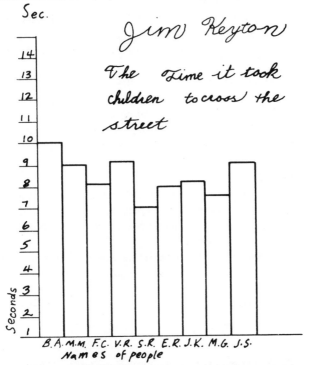

Sec.

Jim Keyton

The Time it took children to cross the street

Seconds

B.A. M.M. F.C. V.R. S.R. E.R. J.K. M.G. J.S.
Names of people

Children are aware from an early age that some pedestrian crossings are more dangerous than others. A child may be permitted to visit a friend only if he uses a specific 'safe' route that avoids a particular intersection. Students are urged to cross streets at locations where traffic police are on duty or where there are WALK lights.

Initial class discussions of the safety of local crossings or of problems children have getting to school will bring out observations children have already made about certain crossings. Children might design and conduct a survey of others in the school to determine a problem intersection.

Motivated by their experiences and suggestions for improvements, children divide into groups to observe traffic and pedestrian flow at designated intersections during different times of day. Data collected may include time intervals between cars arriving at an intersection and the time it takes students to cross the street. If the crossing has a traffic signal, the students may compare the signal time with student crossing times or waiting times to determine the correct timing of the light. Using the data to make tables and graphs, the children can then assess the safety of the intersections and make suggestions for improvements. Classroom simulation of traffic and pedestrian flow is beneficial both as a means of testing data collection methods and as a means of testing proposed improvements.

Students may compare the data collected at a problem intersection with that gathered at a 'safe' crossing. They can then use the results to suggest safety features or traffic controls to improve the hazardous crossing.

As the children collect their data, draw conclusions, and recommend certain improvements, they may see the need for other data such as car speeds, sight distances, and car braking distances. Other activities which the children may find helpful include the investigation of the cost of suggested improvements, the construction of model layouts and model traffic lights, the production of films for the safety education of children and motorists.

A documented written report to the proper authorities may help the children create awareness about a particular problem. In many cases they will make formal presentations or meet informally with officials to achieve a proposed change.

Investigating solutions for improving flow of automobile traffic at a particular intersection may lead students into the Traffic Flow challenge. Promoting driver or pedestrian awareness may lead students into the Advertising challenge.

Exercises

1 Describe how some of the themes mentioned on p. 166 could be developed with (a) eleven-year-olds, (b) fifteen-year-olds.

2 'Like most disorganised armies we (mathematical educators) have our shibboleths, and amongst the most prominent are "real", "useful", "concrete". An examination of what these do and should represent may not be without profit.' (Carson, 1913.) Carry out that examination bearing in mind Carson's point that the pupil's 'reality' is circumscribed and may not correspond to that of the teacher or textbook writer. Provide examples of mismatches.

3 Discuss changes in organisation, etc., which might be necessitated should a school wish to adopt the USMES materials/approach. How might problems of dissemination be eased?

4 What criticisms would you expect a behaviourist to make of the USMES approach (and vice versa)?

5 Are the structuralist and USMES approaches incompatible? Give reasons.

6 Investigate the success (or otherwise) USMES has achieved in effecting changes in school practice. What are the lessons to be learned?

2.   The reform period in Great Britain

The context in which mathematical curricular reforms took place in Britain during the period 1960–75 differed considerably from that in the United States. As we have seen in previous chapters, two characteristics peculiar to the British scene strongly influenced the way in which reforms proceeded.

First, the reforms in Great Britain were closely connected to, and greatly influenced by, the reorganisation of the educational system, a highly political procedure which was carried out by government and administration. The landmarks of the development were the Education Act of 1944, which gave secondary education to all and laid the way for the raising of the school-leaving age to 16 (delayed eventually to 1973), and the Circular of 1965 which requested authorities to prepare and submit plans for reorganising secondary education 'on comprehensive lines'. Developments were also influenced by the reports of the official committees chaired by Crowther (1959), Newsom (1963) and Plowden (1967) which emphasised the need to make changes in school curricula.

Secondly, teachers had a strong say in educational decision-making and, in particular, were seen as competent to deal with questions of content and method. As was shown earlier, it was the curriculum disputes of the nineteenth century that led to the formation of teachers' associations.

The two factors were to some degree in opposition. The growing involvement of the state in education was a threat to the power traditionally exercised by the teachers, who felt called upon to defend their domain. The way in which compromise was achieved is best illustrated by the history of the Schools Council. This had its origin within the Ministry of Education as a Curriculum Study Group. However, the teachers' unions and the local education authorities feared this apparent threat to their autonomy: the outcome was the establishment in 1964 of the Schools Council for the Curriculum and Examinations; a body in which teachers were in a majority. (Some fourteen years later the Council was reconstituted and the teachers' powers diminished.)

The compromise arrangements of 1964 were indeed to become a characteristic of the whole reform era. The state provided the organisational frame within which the substantial work of reform could be executed by teachers. Thus, with the exception of some of the early projects (e.g. SMP, Nuffield Mathematics Project) which were endowed by private enterprises and foundations, the reforms were funded out of public money. In general, however, the money made available for curriculum development in Britain was considerably less than for comparable projects in the USA.

Local teachers' centres were seen as being important agencies for the trial and dissemination of curricular materials and, later, for their development. The centres were conceived as meeting places for the teachers of a town or an area, supplying informal contact as well as courses, consultation and more general, in-service education. They housed collections of all types of teaching materials and provided an additional link between teachers and administration. Their most important task, still largely unfulfilled, was to serve as bases for teacher-centred curriculum development (see pp. 72–3).

By doing this they would help strengthen the key principle that the teacher is not the *consumer* of curriculum development, but the *producer*. Subject specialists and other experts offer advice, but it is the teacher who takes decisions concerning goals, content and methods. The teacher must be enabled to act autonomously and with professional competence. The aim of the centres then is to support continuous cooperation within working groups and between groups and schools, a cooperation which should go further than merely the development of materials.

2.1. *Projects relating to the new-math approach*

The peculiarities of the British situation did not result in conceptions totally different from those found in the USA. However, significant differences of thought can be discerned. Most notably, the British outlook was basically incompatible with the behaviourist approach. As a consideration of tradition and influences would suggest, a preference existed for what we have called the formative and the integrated-teaching approaches. Nevertheless, at the beginning of the reform movement, the most urgent need was generally identified as being the modernisation of syllabuses.

As in other countries, criticism of existing curricula came from industry, commerce and the universities: a public call to reconsider the curricula of secondary schools, particularly in the sixth form, was made in the Crowther Report (1959).

The first mathematics projects of the 1960s had, therefore, a 'New-Math air' about them, although the approach was tempered by British characteristics. The two most important projects of this period were the SMP, started in 1961, and its Scottish complement the SMG, started in

1963. Both of these projects were enormously successful in terms of market penetration. The materials of the SMP were used in over 50% of English secondary schools, while those of the SMG completely dominated in the Scottish academic schools.

Although the SMP was later to diversify, both of these projects were initially concerned with the secondary school curriculum and, in particular, with the course to be followed by the academic grammar-school student, who was to be provided with a better, more up-to-date mathematical education. For this purpose, new topics such as linear algebra and transformation geometry were introduced. Algebraic structures were given more consideration, but no strict axiomatic presentation was envisaged. (Deductive work was, if anything, given less emphasis than previously, when it had appeared in a watered-down version of Euclid.) The main intention was not to prepare students for university, but to introduce them to modern applications of mathematics in a technological society. Thus SMP, which was funded by industry, laid emphasis on a wide variety of applications such as computer mathematics, statistics, probability, operations research (linear programming, critical path analysis, transport flow). However, not all these applications found a permanent place in the school course, for, at secondary level, what is taught still tends to be determined by what is examined rather than by what is possible.

Case study

The following extract is taken from one of the first SMP texts to be published, *Book T4* (1965), intended for 15-year-olds. The SMP books were written by practising (yet particularly gifted) schoolteachers. (The author of this extract later became the head of a university mathematics department.) The content we reproduce is 'New Math', but the style is no longer abstract. However, there is little strikingly novel in the method of presentation: there is no pretence at 'discovery' or openendedness. What is new is the amount of explanation, the sheer number of words as opposed to symbols. It was only with the second-wave of SMP texts *(Books 1–5)* that changes in methodology were emphasised.

1. Vectors

To ask what a vector is, is like asking what a number is; it is something we think we know all about, until we are asked to give a definition, when we suddenly discover we know nothing at all precise about it.

Either we have to give a very sophisticated definition, or else we find ourselves taking something to go on with and listing the properties which we want it to have.

This is a situation which frequently arises in mathematics. What, for example, is a straight line? To say it is the shortest distance between two points suggests ideas of stretched threads and light rays and is a good description to be getting on with; but it can hardly be called a definition, since it involves awkward ideas like 'distance' and 'short', and, what is even worse, the existence of something which can be described as 'shortest'. What we really need to know about straight lines are their *properties*; that there is just one of them joining any two points and that two of them (in the same plane) will meet unless they are parallel. Let us therefore take the plunge, remembering that we may need to come back later and take a more sophisticated look at what a vector is.

1.1. **Row and column vectors**

A vector is a single object which needs several numbers to describe it; it is therefore given by a list of numbers, called its *components*. These can be written either as a *row* or as a *column,* for example:

$$(2 \quad 3 \quad 0) \quad \text{or} \quad \begin{pmatrix} 2 \\ 3 \\ 0 \end{pmatrix}.$$

We call the first a *row vector* and the second a *column vector*; we do not consider these the same: it is sometimes convenient to have the two kinds. The *order* of the components matters; $(2 \ 3 \ 0)$ is not the same vector as $(2 \ 0 \ 3)$. To a geometer or an applied mathematician or a physicist the numbers will usually be distances or coordinates, so that a vector is a sort of instruction to travel so far in a definite direction: $(2 \ -1 \ 0)$ would mean 'go 2 units in the x-direction, 1 unit in the negative y-direction, and 0 units in the z-direction'. To an algebraist a vector can mean this too, but it can also be *any* list of numbers. $(2 \ 3 \ 0)$ could mean '2 eggs, 3 apples and no bananas' or '2 men, 3 women, and no children' or '2 soups, 3 salmon mayonnaise and no peach melba' or anything else that a mathematician likes to make it refer to. Bertrand Russell once defined pure mathematics as 'the subject in which we never know what we are talking about' (and he added, 'nor whether what we are saying is true', but that is another story!). A vector is thus a particular kind of matrix; a row vector with n components is a $1 \times n$ matrix; a column vector with n components is an $n \times 1$ matrix.

1.2. **Sums**

Suppose I go into a shop and ask for 3 apples, 5 oranges and 4 bananas. My purchase can be described as (3 5 4). If now someone else comes in and buys 4 apples, 2 oranges and 1 banana, his purchase is (4 2 1). The total is obviously 7 apples, 7 oranges and 5 bananas, that is, (7 7 5). We therefore write

$$(3 \quad 5 \quad 4) + (4 \quad 2 \quad 1) = (7 \quad 7 \quad 5).$$

We call the individual members the components of the vector and we say the sum of two vectors is the vector whose components are the sums of their respective components. A vector is a sort of inventory, that is all. We treat column vectors in the same way.

We define

$$(u_1 \, u_2 \, u_3) + (v_1 \, v_2 \, v_3) = (u_1 + v_1 \; u_2 + v_2 \; u_3 + v_3),$$

$$\begin{pmatrix} u_1 \\ u_2 \\ u_3 \end{pmatrix} + \begin{pmatrix} v_1 \\ v_2 \\ v_3 \end{pmatrix} = \begin{pmatrix} u_1 + v_1 \\ u_2 + v_2 \\ u_3 + v_3 \end{pmatrix}.$$

We never add a row and a column vector together.

1.3. **Multiplication by a single number**

If two people go into a shop and each asks for the same purchase vector of 3 apples, 5 oranges and 4 bananas, then it is obviously sensible to say that their purchase is $2 \times (3\ 5\ 4) = (6\ 10\ 8)$. In words, a vector is multiplied by a number when every component is multiplied by that number.

$$k \times (u_1 \, u_2 \, u_3) = \ldots?$$

and

$$k \times \begin{pmatrix} u_1 \\ u_2 \\ u_3 \end{pmatrix} = \ldots?$$

This property and that of Section 1.2. are the essential properties that vectors must have.

  (a) If **a** is the vector $(a_1 \, a_2 \, a_3 \ldots)$ and **b** is the vector $(b_1 \, b_2 \, b_3 \ldots)$, then **a** + **b** is the vector

$$(a_1 + b_1 \quad a_2 + b_2 \quad a_3 + b_3 \; \ldots).$$

  (b) If k is a number, then $k\mathbf{a}$ is the vector

$$(ka_1 \quad ka_2 \quad ka_3 \; \ldots).$$

There can be any number of components, but of course there will always be the same number in any set of vectors we are talking about. Also, we may write our vectors either as rows, or as columns, so long as we are consistent in writing all vectors of the same kind in the same way.

Exercise A

1. A kit for a radio tuner contains 3 valves, 12 resistors, 10 capacitors and 4 coils. Express this as a vector and show how to obtain the requirements for 50 kits.
   (SMP *Book T4*, pp. 113–115.)

Exercises

1 Write an essay on the Schools Council's contribution to mathematical education.
2 Investigate the early attempts made by SMP to introduce algebraic structure to O-level pupils. (Consult early drafts, *Book T4,* etc.) What lessons are there to be learned?
3 Study the way in which SMP and other projects came to place increasing emphasis on teaching methods. What were the reasons for this?
4 Compare and contrast the SMP paragraphs on 'definitions' with the views of Branford and Davis (see p. 165). See also how vectors were introduced and defined by 'New-Math' authors such as Papy. What are the advantages and disadvantages of the various approaches?
5 Are you convinced by the sudden appearance of, and distinction between, row and column vectors? If not, how could one make improvements? Provide further examples in mathematics teaching where we tend to make apparently arbitrary distinctions, so that 'later' our notation and symbolism will conform to advanced usage.
6 What can be inferred from the extract about the teaching methods it is hoped to encourage?

2.2. *Projects based on the formative approach*

As in the USA, the emphasis in curriculum development soon shifted from subject-orientation to personality-orientation. In Britain, however, this change took place in a different way. In the USA it was the theoretical acceptance of cognitive psychology which inspired, mainly due to Bruner, a fusing of the two aspects: from this the structuralist approach resulted. In Britain, the move to a consideration of psychological problems owed little to theoretical deliberations, but was rather a result of teacher-involvement in curriculum development.

The more they were concerned with curricular reform, the more their long-standing, practical interest in child development reappeared. The structuralist approach although recognised (mainly due to the efforts of Dienes) did not take root, the behaviourist with its claims for a 'teacher-proof' course was condemned; instead, a formative approach emerged as a natural foundation on which to base curriculum development.

Another reason for the appeal of the formative approach was that the introduction of modern mathematical content never gave British syllabuses the rigidity it stamped on syllabuses elsewhere. 'The British place more emphasis than Americans on a method of instruction appropriate for the maturity level of the student and less emphasis on the explicit display of the mathematical structure being studied' (van Engen, 1973). This is the remark of an American observer about the SMP. From the extensive consideration of mathematical applications (not all very useful in themselves from the standpoint of pure science), it was not so great a step to a freer exploration of the pupils' environment: a favourite context for the learning of mathematics in the new projects.

The most important project relating to the formative approach was the Nuffield Mathematics Project (1964), whose suggestions were followed by a great number of primary schools inside and outside Britain.

By seeking to create stimulating learning situations in which mathematics serves as an instrument for problem solving, and encouraging communication in which mathematics proves a useful and precise language for understanding and explanation, the Nuffield Project displayed its affinity with projects such as the Madison.

The differences between Nuffield and Madison reflect, perhaps, more the differences in the two educational systems in which they arose, than differences in aims. The open curriculum in Britain presupposes and permits classroom-based curriculum development: the involvement of teachers encourages a pupil-centred curriculum.

Exercises

1 'Every child should experience the joy of discovery.' (Whitehead, 1929.) Discuss.

2 The primary school age is 'a time when children explore tirelessly both their own powers and all that is of interest in their environment'. (*Planning the Programme*, HMSO, 1953.) Comment. What are the consequences for mathematics teaching?

3　Compare and contrast the approaches to mathematics learning exemplified in, on the one hand, the Nuffield guides *Pictorial Representation* and *Environmental Geometry* and, on the other, the various guides on *Computation and Structure*. What questions would a teacher have to resolve before such work could be adopted in the classroom?

4　It could be argued that the Nuffield guides often leave children with ham-fisted, 'self-devised' techniques for carrying out numerical tasks such as, say, ordering fractions. Is this a fair criticism? What are the implications for mathematical education of such a policy?

2.3.　*Projects based on the integrated-teaching approach*

Although we have spoken of the affinity of Nuffield with Madison, the former would really seem to lie somewhere between Madison and USMES. This is apparent both in the form in which its materials were produced and in its choice and the complexity of the learning situations. The relation between the formative and integrated-learning approaches is rather blurred in Britain, particularly in the primary school. (This, then, is perhaps a good opportunity to remind the reader of the purely heuristic interest the authors have in distinguishing between the various approaches!)

Nevertheless, a third generation of projects may be discerned as originating from the activities of the Schools Council. Initially, these were responses to political and administrative changes.

Secondary education for all had originally been provided in a tripartite form: grammar schools (for the academic), technical schools (in limited numbers, and with a vocational orientation), and modern schools (for the majority of children). For a variety of reasons the mathematics provided in the last named was frequently restricted and impoverished, often going little beyond arithmetic and mensuration. Mathematics (as distinct from arithmetic) was seen as a selective school discipline. Fired by the move towards comprehensive schools, there was now greater pressure to provide a significant mathematics course for all students. All pupils should be made aware that mathematics provided qualifications which would be useful and necessary for them in their future lives, both in employment and at home; mathematics offered a basis on which rational decisions could be made in a variety of spheres. It was in the 'best' tradition of modern-school thought, as evidenced in the reports of Hadow (1926) and the Mathematical Association (1959), that pupils of average and below-average ability would respond most readily to an approach to mathematics based on practical and social considerations, strongly integrated to other subjects in their curriculum.

A second, organisational spur to further curricular initiatives was provided by the decision to extend compulsory schooling for an additional year up to the age of 16. What had previously been a four-year course for the 'early school-leaver' had now to be redesigned to cover five years.

Several projects were initiated by the Schools Council, all of which shared common traits including an attempt to integrate a variety of features, such as mathematics, the environment, physics, social studies, and language. It suffices here to mention Mathematics for the Majority (MMP), Science for Young School Leavers, Integrated (Nuffield Combined) Science and the Humanities Curriculum Project.

The MMP, which commenced work in 1967, chose to model its organisation on the Nuffield Project, providing only teachers' guides. That it failed to have any marked effect on practice in schools was largely as a result of this, and of other managerial decisions which are not the direct concern of this chapter. The need for classroom materials was realised and a 'Continuation' project (MMCP) to provide these was established in 1971. This project could be even more clearly identified with the integrated-teaching approach, as is clear from the titles of its publications. These are packs which include work cards, booklets, models, tapes, games, puzzles, etc. The titles so far published are 'Buildings', 'Communications', 'Travel' and 'Physical Recreation'. (Eight other packs were prepared but at present there are no plans to publish them.)

In general, the materials produced by these projects were more modest than those of similar projects in the USA. They were somewhat less professional in their presentation, but compensated for this by their greater air of spontaneity and adventurousness.

Case study

We can best indicate the variety of mathematics covered in an MMCP pack by reproducing* the contents pages of one of them – *Physical Recreation*. It is instructive to see how a large variety of recreational activities can provide stimuli for the study of mathematical topics, concepts and methods.

---

* Reproduced by permission of the Schools Council Publication and Schofield and Sims Ltd.

PHYSICAL RECREATION

Title	Topic	Maths content
1 Running track	Going round in circles; measuring circles; running tracks.	Circumference and diameter of circles; circles as a locus, the spiral (rather a complicated one); computation and measurement.
2 Breathing rates	How big is a litre? Air in your lungs.	Large numbers, mensuration, especially volumes in metre units; model-making.
3 Diving	Comparing three dives.	The parabola (an interesting example of a distance time graph).
4 Competitive swimming	Timing a swim; scoring at a swimming gala.	Averages, running totals, estimation and precise measurement.
5 It's a goal!	The soccer pitch; scoring goals; shooting at goal.	Co-ordinates, perspective, angle subtended at a point, angles in same segment.
6 Matrix soccer	Playing the game; matrix soccer.	Relating an ordered pair in matrix form to a vector displacement; use of negative numbers within left/right, up/down convention.
7 Does it float?	Things which float; what causes an object to sink; what sizes of boats are needed for different weights.	Volumes of cuboids, Archimedes' Principle, ideas of density.
8 Sailing	Sailing; courses; bearings.	Bearings and angles.
9 Mountain walks	Contours; a walk in lakeland; heights of mountains.	Applying a formula, scale, slope. Contours could lead to topological notions, e.g. continuity and discontinuity.
10 More about mountains	A model mountain; a lakeland scene; distances.	Interpreting contours, model-making, the cone and the pyramid.
11 Wind	Wind speed; anemometer.	Interpretations of regions on a graph, angular velocity, rotational symmetry.
12 Ropes	Breaking strength; stretching; taking the strain.	Friction, tension in a string (quantitative); Algebra of permutations (plenty of follow-up here – for example,

Title	Topic	Maths content
		in 'placing in order' competitions); transitivity through the order relations $>$ and $<$. Thus $A > B$ and $B > C$ $\Rightarrow A > C$. Note that the order relations are **not** equivalence relations as it is not true that either $A > A$ or that $A > B \Rightarrow B > A$.
13 Strength and direction	Strength to weight ratios; directions; map references.	Ratios, measurement in kilograms, angles, bearings, co-ordinate systems (in particular, map references).
14 Gone fishing	A fishing game.	Fractions and percentages.
15 Fishing lines	Types of fishing lines; fishing techniques.	Units of measurement, diameter of circles, fractional and decimal parts; reading a graph, functional relationship, Archimedes' Principle.

Exercises

1 Investigate the evidence for the assumption 'that pupils of average and below-average ability . . . respond most readily to an approach to mathematics based on practical and social considerations, strongly integrated to other subjects in the curriculum'.

2 Describe some of the problems arising from a decision to extend the duration of compulsory education.

3 Take some of the MMCP 'titles' and explain in detail how one could derive the 'maths content' from the given situations.

4 Despite the richness of the MMCP packs, their sales have been disappointing. Give some possible reasons for this and discuss how the problems of dissemination might have been eased.

5 Take an MMP guide and try to translate a chapter or section of it into classroom terms (i.e. through the preparation of pupils' materials, 'starting points', etc.). *(Machines, Mechanisms and Mathematics* offers some particularly novel ideas.)

6 Discuss the problems which arise from trying concurrently to provide both prevocational and general education and compare the ways in which the various projects mentioned in this chapter have attempted to reconcile these two aims.

7

Evaluation within Curriculum Development

Evaluation is ubiquitous and eternal: people have been judging the worth of their own or others' handiwork since the time of Genesis. In education, every teacher evaluates his teaching in one way or another, if only to help him decide what to do next. Teachers themselves are evaluated by their employers and also by their students, who at least since the time of the medieval university have 'voted with their feet' – when they could – to express their satisfaction or dissatisfaction with the instruction they were receiving. But just as the curriculum had not been widely perceived as something to be 'developed' until the middle of this century, with the advent of curriculum development projects, so the need for an explicit, formal evaluation of the curriculum did not arise until people began to ask whether the projects had been worthwhile. The job of curriculum evaluator, like the job of curriculum developer, is a twentieth-century invention.

1.   **The process of evaluation**
        What is evaluation? The basic idea is simple: evaluation is the process of judging the value or worth of something. To evaluate, one needs an object, a scale of value, and some means of gathering information about the object so that the scale of value can be applied to the information.
   For example, in evaluating the value of a house prior to its sale, the house is the object, money provides the scale of value, and the information to be considered concerns such matters as the purchase price of the house, the value of improvements made to it, and recent selling prices of comparable houses, as well as intangibles such as its likely appeal to prospective buyers. Even in this simple example, one

can see that the most difficult part of evaluation is not the technical issue of how to gather the information; it is deciding what kind of information is relevant and how information from different sources should be weighted and combined. As Jacobsen (1978, p. 3) put it: 'Basically evaluation consists of three activities: asking significant questions, gathering information to answer the questions, and interpreting the results'. In Jacobsen's terms, the hard part concerns the value judgements needed to decide which questions are significant and how the results should be interpreted.

Curriculum evaluation is especially difficult for two reasons. The first concerns the nature of the curriculum. A curriculum is not a tangible object like a house. It does not exist as an entity that can be made to stay put while it is measured, photographed, and judged. Instead, a curriculum is an abstraction that can only be glimpsed through such means as the analysis of statements of aims, the observation of content actually taught, and the assessment of what pupils have learned. Two classrooms in which the same curriculum is supposedly being implemented may be quite dissimilar when one looks at what the teachers and children are doing, what is being taught, and what is being learned. Any attempt to evaluate a curriculum must somehow deal with its 'situation-specific' nature: the curriculum is manifested differently at different times and in different places. How can one capture it for evaluation?

The second reason curriculum evaluation is difficult arises from the socio-political context in which educational decisions are made. Curriculum evaluations are ordinarily not undertaken out of dispassionate intellectual curiosity; rather they are undertaken because decisions must be made. Should a new curriculum replace the old one? Which of two curricula would be more effective with mixed-ability classes? Should this curriculum development project continue to receive government funds to support its work?

The closer one approaches the curriculum to be evaluated, the more aware one becomes of its manifold qualities. Some of its features may be quite good, while others may need improvement. In some situations it seems to be working well; in others not at all. It is likely to be a very mixed bag of the good, the bad, and the questionable.

To make a decision about a curriculum – whether or not to adopt, modify, or abandon it – one needs to step back from it so that its manifold qualities can merge and permit the assessment of a single index of its worth. How good is it? Is it better than the alternatives? Does it give value for money? The educational decision-maker can answer such questions only by combining the many features of the curriculum, ignoring the details and giving the greatest weight to those features deemed most important. Curriculum developers, who naturally are close to the curriculum they have developed, usually see it from a different perspective than that of the decision maker. Curriculum evaluation is ordinarily done at the behest of the decision maker, but it cannot ignore the concerns of the other participants in the evaluation process.

Curriculum evaluation is an interactive psychological and socio-political process. It is interactive because it involves the knowledge, values, and beliefs of those whose work is evaluated, those who do the evaluating, and those for whom the evaluation is done. Each participant in the evaluation process has his own perspective on it, and each influences the other participants. Curriculum evaluation is a psychological process because it affects peoples' beliefs about their work and themselves. It is a socio-political process because it affects decisions as to what shall be taught to whom. In democratic societies, curriculum evaluation should itself be democratic in the sense that those affected by decisions about the curriculum should participate in evaluating the curriculum and making the decisions.

Any curriculum evaluation effort should be preceded by an attempt to clarify the purposes to be served by the evaluation. One should identify the decisions to be made and the actions to be taken on the basis of the evaluation. A curriculum evaluation should not be undertaken simply to legitimate a decision that has already been made on other grounds.

The curriculum evaluator is, in a sense, caught in the middle between those on the 'firing line' and those at the 'headquarters'.

In the classroom one sees the uniqueness of the situation and the many factors that can influence teaching and learning. Further, one can respect the individual pupil's right to develop in a unique way under the influence of a curriculum. In the Ministry of Education or the local authority's offices, however, one sees that society requires certain levels of mathematical competence from its citizens and that curricula must be chosen so as to maximise this competence without denying teachers their right to tailor the curriculum to fit particular circumstances. The rights of individuals and society must be respected and brought into balance. The effective evaluator is one who sees the multiple perspectives of the participants in the evaluation process and who is able to reconcile the concerns of teachers and administrators.

Exercises

1 An evaluation can be of interest to many people: to the initiators of a project, to the funders, to the writing teams, to heads of schools, and to individual teachers (whether or not they are using the project's materials). Choose a project well-known to you and list questions concerning it which each of these groups might wish to have answered. How could you attempt to provide those answers? In what ways would your findings best be communicated to the interested parties?

Bear this particular exercise in mind whilst reading the case studies in this chapter and consider in each instance whose questions the evaluator has chosen to answer, and the methods of communication used.

Repeat the exercise after you have finished reading the chapter.

2 To what extent is it desirable/essential that innovators and evaluators should share the same criteria? What consequences are likely if (a) criteria are completely common, (b) criteria are incompatible?

Bear these questions in mind when reading the case studies in this chapter.

2. Metaphors for evaluation

What is done in a curriculum evaluation depends on how the process is conceived, and curriculum evaluators have approached their tasks from various directions. The dominant approach to curriculum development and evaluation has been to conceive of them as engineering processes. In this metaphor, the school is a factory and education is a production process. Pupils enter as raw material and emerge at the end as finished products. The curriculum is seen as an instrument for converting raw material into finished product. Curriculum evaluation in this metaphor requires a list of specifications, just as one would evaluate a refrigerator or an automobile.

As Scriven (1967) has pointed out, there are two contrasting approaches to evaluating a 'teaching instrument'. The first is 'intrinsic' evaluation in which one appraises the instrument itself: what are its qualities and characteristics? The second is 'pay-off' evaluation in which one looks at the effects of the instrument on the pupil. Scriven (1967, p. 53) contrasts the approaches with a homely illustration: 'If you want to evaluate a tool, say an axe, you might study the design of the bit, the weight distribution, the steel alloy used, the grade of hickory in the handle, etc., or you might just study the kind and speed of the cuts it makes in the hands of a good axeman'. Scriven's characterisation of curricula as educational instruments shows his adherence to the metaphor.

Education as production; curriculum development as the design and construction of the machinery for production; examinations and assessment as quality control; curriculum evaluation as the monitoring and maintenance of the machinery – these are expressions of a metaphor that has pervaded educational thought in the USA and that has spread around the world as other countries have adopted and adapted US practices. The metaphor has severe limitations, however, and these seem especially serious with respect to evaluation. Curriculum evaluators who have viewed their role as a kind of engineering have found that the information they gather about the educational 'product' and the educational 'machinery' is typically too full of error, fragmentary, late, and irrelevant to be of much use in decision making. A large number of evaluation studies have been conducted that have had negligible impact on decisions regarding the curricula being evaluated.

Frustrated by their impotence and aware that the engineering approach too easily sidesteps questions of conflicting and immeasurable goals, some evaluators have sought other metaphors in which to cast their work. For example, the medical model of evaluation (Anderson, Ball, Murphy, and associates, 1975) attempts to go beyond the engineering approach by viewing the curriculum as a treatment and the pupil as a patient whose performance is affected by many interacting factors. In using the medical model one looks for unintended side effects of the treatment and not just those intended by the curriculum developer. Also, one does not conclude evaluation at the end of the treatment. Rather, one conducts follow-up investigations to see what the long-term outcomes might be. The medical model treats the pupils not as raw material to be moulded into shape, but as organisms whose behaviour needs to be understood. One does not stop at assessing the magnitude of the treatment's effects; one tries to understand the processes that produce the effects.

Another metaphor for curriculum evaluation is that of journalism. The evaluator becomes a reporter, tracking down leads and using interviews and observations to put together a terse and timely report. The evaluation by Stake & Gjerde (1974) of the Twin-City Institute for Talented Youth fits this rubric. Stake (1967, 1977) has argued for evaluation as *portrayal* rather than as *analysis*. He uses the methods of participant observation from sociology, and ethnographic fieldwork from anthropology, to assemble a collage that will 'tell the story' of the curriculum to a variety of audiences. As Jenkins (1976, p. 41) notes, 'Full portrayals will probably exhibit similarities to literary criticism, film documentary, historical research, law and clinical psychiatry.' Stake's concern is to provide a comprehensive portrayal rather than a focused sketch, so perhaps some of these other metaphors are closer to his intentions than is journalism. Parlett & Hamilton (1977) propose a view of evaluation as 'illumination' that resembles Stake's approach but that rejects more decisively what they see as the 'agricultural-botany paradigm' popular in psychological research and in educational evaluation studies. They want to use instead the 'social-anthropological paradigm' that takes account of the contexts in which the curriculum functions. Such proposals, then, basically represent a rejection of the engineering (or agricultural-botany) metaphor and are attempts to broaden the evaluator's view of the curriculum as it is variously manifested.

Alternative evaluation methodologies have been much discussed in recent years, but few have been applied to mathematics curriculum development projects, although some evaluation studies in the last few years have used techniques borrowed from these methodologies. Evaluators of mathematics curricula have, indeed, been too eclectic in their approaches to permit a neat classification. Consequently, in the following sections, the approaches to curriculum development identified in chapter 5 will be used to structure the discussion. An examination of various evaluations of curriculum development projects may illuminate some of the points made above and will certainly raise some new points for consideration.

3.   **The evaluation of projects based on the new-math approach**

Since the 'New-Math' curriculum development projects of the 1950s and 1960s were primarily concerned with updating the content of the mathematics curriculum, it is not surprising that evaluations of these projects tended to be directed at the question of content: could pupils learn the new content and still learn the old content too? When the question is stated this boldly, it is clear that something has to give, unless one views either the old content as subsumed by the new or the pupils as capable of learning content they have not been taught. Some of the 'New-Math' reformers appeared to hold both of these views. That is, they argued that pupils did not need great amounts of practice in, say, multiplying fractions or factoring algebraic expressions. They considered these to be trivial skills that would be learned incidentally as pupils worked more challenging exercises and problems that made use of the skills. Other reformers simply took the view that the loss of such content was not serious and that pupils could get along quite well without it; the important thing was for the pupils to understand the basic ideas of modern mathematics.

Consequently, in the first formal evaluation studies of New-Math curricula a standardised mathematics achievement test was given to a group of pupils who had used the new curriculum and to a roughly comparable group who had not. The curriculum developers would also construct and administer a test designed to assess the novel content of their programme. This test was usually given to both groups, although sometimes it was so clearly inappropriate for the 'traditional' group, because of such features as unusual notation and terminology, that it would be given to the 'experimental' group only.

The results of such studies were about what one would expect. (See Begle & Wilson, 1970, for a review of some studies done in the USA.) The experimental group did either about as well as, or somewhat worse than, the traditional group on the standardised test, and substantially better on the specially constructed test. This sort of information reassured those who worried about possible detrimental effects the new curricula might be having, although some still worried about the deficiencies in the standardised test performance.

The reformers soon realised, however, that such evaluation studies were not very helpful in distinguishing between the effective and ineffective parts of the new programmes. Standardised achievement tests are poor devices for such diagnosis; they are designed to discriminate students' performance, they ordinarily yield a single score, and they do no more than sample curriculum topics. Further, they tend to emphasise the recall of factual information and the performance of routine calculations.

Some of the new reformers also concluded that these early studies, which typically examined achievement over one school term or year, did not give the new curricula an opportunity to show their long-term effects. Although the sixth-grade pupils in a modern programme might lag behind their counterparts in traditional programmes in their ability to compute with fractions and decimals, say, the reformers thought that the 'modern' pupils might 'catch up' during the next few years owing to their superior understanding of the concepts involved.

3.1. *The National Longitudinal Study of Mathematical Abilities*

To provide a more comprehensive and extended evaluation of pupils' performance in the New-Math curriculum programmes in the USA, the School Mathematics Study Group began in 1961 the National Longitudinal Study of Mathematical Abilities (NLSMA). NLSMA was intended not simply to yield detailed information about the achievement of pupils following modern and traditional curricula, but also to investigate the growth of mathematical skills and abilities.

The designers of NLSMA believed that achievement in mathematics is multifaceted and that its assessment required a battery of short tests, or scales, aimed at different facets. NLSMA used a matrix model for classifying the items in the scales according to two dimensions. One dimension was the mathematical content with which the items were concerned. This dimension was relatively unambiguous: the common topics from syllabuses for each course provided the scheme for organising the content, and items could be classified according to content without much difficulty. The other dimension, however, posed more of a problem. It dealt with the 'cognitive process' that was presumably tapped by an item. Recall that standardised mathematics achievement tests were deemed inadequate for assessing achievement because they relied so heavily on the pupil's memory for facts and procedures and so little on such things as his understanding of why the procedures worked and his ability to apply his knowledge of mathematics to new situations. The designers of NLSMA wanted to tap some of these 'higher cognitive processes'. In setting up the second dimension of the model, they borrowed and adapted the taxonomy developed by Benjamin Bloom (1956) and his colleagues to aid in devising objective test questions to measure different types of instructional objectives. Bloom's taxonomy was not set up with any one subject field in mind, and it does not seem to fit mathematics very well. Numerous alternative taxonomies have been proposed for mathematics (see Wilson, 1971, for some examples), and one of the earliest of these was that devised for NLSMA. The following excerpt* is from Begle & Wilson's discussion of the NLSMA model (1970, pp. 373–4). Both the mathematical content and the cognitive process dimensions have been simplified for use in the discussion of the results of NLSMA.

> The essential idea of the model is that measures of mathematics achievement, or test items, or objectives of mathematics instruction, can be classified in two ways: (*a*) by categories of mathematical

---

* Reproduced by permission of the National Society for the Study of Education, Chicago.

content, and (*b*) by levels of behavior. The levels of behavior reflect the cognitive complexity of a task (*not* simply the difficulty of a task). In the model presented in Figure 1, the categories of mathematical content are number systems, geometry, algebra. The levels of behavior are computation, comprehension, application, and analysis.

The model, as it is presented, requires explicit specification of the terms along each dimension. These specifications are as follows:

CATEGORIES OF MATHEMATICS CONTENT

Number Systems   Items concerned with the nature and properties of whole numbers, integers, the rational numbers, the real numbers, and the complex numbers; the techniques and properties of the arithmetic operations.

Geometry   Items concerned with linear and angular measurement, area, and volume; points, lines, planes; polygons and circles; solids; congruence and similarity; construction; graphs and coordinate geometry; formal proofs; and spatial visualization.

Algebra   Items concerned with open sentences; algebraic expressions; factoring; solution of equations and inequalities; systems of equations; algebraic and transcendental functions; graphing of functions and solution sets; theory of equations; and trigonometry.

LEVELS OF BEHAVIOR

Computation   Items designed to require straightforward manipulation of problem elements according to rules the students presumably have learned. Emphasis is upon performing operations and not upon deciding which operations are appropriate.

Comprehension   Items designed to require either recall of concepts and generalizations, or transformation of problem elements from one mode to another. Emphasis is upon demonstrating understanding of concepts and their relationships, and not upon using concepts to produce a solution.

Fig. 1. A model for mathematics achievement

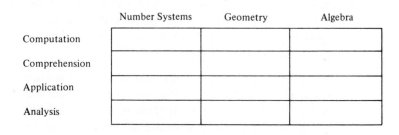

	Number Systems	Geometry	Algebra
Computation			
Comprehension			
Application			
Analysis			

Application Items designed to require (*a*) recall of relevant
knowledge, (*b*) selection of appropriate operations, and (*c*)
performance of the operations. Items are of a routine nature. They
require the student to use concepts in a specific context and in a way
he has presumably practiced.

Analysis Items designed to require a nonroutine application of
concepts. The items may require the detection of relationships, the
finding of patterns, and the organization and use of concepts and
operations in a nonpracticed context.

The second dimension, levels of behavior, is both hierarchical and
ordered. It is *ordered* in the sense that analysis is more cognitively
complex than application, which is in turn more cognitively complex
than comprehension, and the computation level includes those items
which are the least cognitively complex. It is *hierarchical* in that an
item at the application level may require both comprehension-level
skills (selection of appropriate operations) and computation-level skills
(performance of an operation).

The use of the model usually requires some rules of thumb. For
instance, outcomes are classified at the highest cognitive level even
though lower level behaviors are required. . . .

NLSMA was one of the largest curriculum evaluation studies ever
undertaken. School systems all over the USA were asked to participate
in a five-year programme of testing. Over 110 000 pupils from 1500
schools in 40 states took part. The schools were recruited on the
understanding that they would not be required or expected to make any
curriculum changes because of the study. SMSG provided no special
services or materials to the schools as a condition of participation.
Unfortunately, the selection of schools and pupils was neither random
nor representative of the US system as a whole. Participation had to be
negotiated and cajoled rather than commanded. Consequently, sub-
urban schools, schools in the Far West of the US, and schools using
SMSG textbooks were over-represented in NLSMA, and the NLSMA
pupils tended to be better than average in mental ability, mathematics
achievement, and socio-economic status.

Three 'populations' of pupils participated in the NLSMA testing
programme, which began in the fall of 1962 and continued until the
spring of 1967. Tests were given twice a year: fall and spring. The
X-population began the study as fourth graders and were eighth graders
in the final year. The Y-population began as seventh graders and were
eleventh graders in the final year. The Z-population began as tenth

graders and were tested for three years until the end of secondary school (grade twelve) and then followed by means of questionnaires for two more years. This design of three cohorts followed for five years, with overlapping at grades 7 and 8 and grades 10 and 11, yielded extensive data on mathematics achievement for grades 4 through 12, as well as the opportunity to study achievement in the same grade at different times (for example, grade 7 in 1962–63 for the Y-population and 1965–66 for the X-population) and in the same cohort over five years.

The mathematics tests for NLSMA were constructed by subdividing the column headings of the matrix model into topics appropriate for the curriculum of a given grade; selecting various cells in the matrix to give variety in both content and cognitive level; borrowing, adapting, or creating items varying in difficulty that were suitable to each cell; trying out the items with another group of pupils; and collecting the items into scales for administration to the NLSMA pupils.

The designers of NLSMA assumed that mathematics achievement depends on more than cognitive abilities. Scales were developed and administered to measure various aspects of attitude toward mathematics, anxiety, achievement motivation, and self-concept. Information was also gathered by questionnaire regarding characteristics of the pupils' teachers, school and community. Begle & Wilson (1970) give further details on the organisation of the study.

NLSMA was undertaken on the assumption that school mathematics curricula were changing rapidly and that 'traditional' curricula might soon become almost obsolete. The study was begun, therefore, in some haste. A complete list of questions to be answered by it was not formulated in advance, and the tests were constructed as the study proceeded, sometimes with little time to spare. Since not all relevant variables could be specified beforehand, the NLSMA population was chosen to be large enough so that 'the effects of unmeasured variables could hopefully be randomized out' (Begle & Wilson, 1970, p. 389). This hope proved to be optimistic at best. When certain variables were measured and samples of NLSMA pupils were selected according to these measures, the samples frequently turned out to be too small to permit reliable analysis. Problems of missing data are endemic in any longitudinal study; in NLSMA they combined with the non-representativeness of the NLSMA populations to frustrate a good many attempts to dip into the data pool and come up with something solid.

When NLSMA was planned, it was intended that the data pool would be used not only for comparing the achievement of pupils in various curriculum programmes, but also for basic research in mathematics education. Once the data had been collected, however, the magnitude and complexity of the data pool and the limitations of time and other resources prevented extensive analyses beyond the comparison of the achievement of groups of pupils who had used various textbooks. In NLSMA, the curriculum was defined by the major textbook that the teacher claimed had been used with the pupils, and most of the analyses carried out concerned the comparison of groups defined by the textbook they had used. Although the comparison of textbook groups was not the sole, or originally the primary, purpose of NLSMA, it became by default the centrepiece of the data analysis. Considerable technical skill and effort went into the textbook comparison analyses; it is unfortunate that similar skill and effort were not applied to other questions which the NLSMA data might have helped to answer.

NLSMA yielded some interesting results regarding patterns of achievement related to the use of various textbooks. Thus, for example, it was shown that when computational ability was tested at the end of grade 8, pupils who had used a conventional textbook tended to perform better than any of the groups using modern textbook series. Yet when the pupils' understanding of number properties was measured, the modern textbook groups did relatively well, and it was the conventional textbook group whose performance was relatively low (McLeod & Kilpatrick, 1969). This sort of pattern, although seldom so pronounced, was characteristic of scales classified in the comprehension, application and analysis rows of the matrix.

Other attempts to compare the various textbook groups made use of the statistical technique known as discriminant analysis (see McLeod & Kilpatrick, 1969). This showed not only that, in a defined sense, the conventional textbook group was well separated from the others, but that there were some clear differences among the modern textbook groups also.

The NLSMA results, then, showed that, contrary to some expectations, pupils who had used modern textbooks did not achieve superior levels of performance in computation over several years while using textbooks that did not stress computation. Instead, the pupils using 'traditional' textbooks tended to do better on the computational scales, and worse on the other kinds of scales, than those using modern textbooks. To oversimplify, pupils tended to learn what was emphasised in the textbooks they used and not something else. NLSMA also showed that modern textbooks were not a homogeneous mass. They seemed to differ considerably in their effects on what students learned, although the NLSMA writers were usually at a loss to explain the source of the differences. There also appeared to be a tendency for the number of differences between textbook groups to decline from grades 4 to 12. This may well have been due to a greater adherence to the textbook by teachers at lower levels. Other factors, such as a greater inadequacy of the NLSMA scales for measuring achievement in the higher grades, may also have contributed to the reduction in differences. (A comprehensive review of the *NLSMA Reports* can be found in Osborne, 1975, where several reviewers comment on the complexity of the analysis and the fallibility of the data.)

NLSMA was unique; such a study will never be attempted again. The circumstances that brought it about – the National Science Foundation's faith in a project's ability to evaluate its own work, the willingness of hundreds of schools to participate in an extensive study that promised little in the way of return to them, the availability of millions of dollars for unfocused and largely exploratory research, the belief that the effects of curricula could be profiled so that consumers could then make rational choices among them – these are unlikely to occur again. NLSMA explored the important notion that mathematics achievement is multi-dimensional, and the NLSMA textbook comparison analyses provided some important conclusions about the association between textbook usage and patterns of achievement. Many of the instruments devised for NLSMA, such as the scales for measuring the ability to apply mathematical concepts in a nonroutine context and the scales for measuring attitudes toward various facets of mathematics, are a rich source of ideas for teachers and researchers.

NLSMA could have been planned and executed with greater care and certainly at less cost. But in a sense the tide had to be taken at the flood or not at all. The moment was seized. Was two and a half million dollars too much to pay to evaluate the work of a project (SMSG) that cost over fourteen million dollars; as well as the work of several smaller projects? A better question is how the evaluation dollars might have been spent in a more efficient way. Some of the most important lessons NLSMA taught were that one cannot correct for problems of planning and design by using a large sample and elaborate statistical techniques, that no study is large enough to answer more than a limited number of questions, and that the biggest temptation in a longitudinal study is to spend so much of one's resources gathering and organising the data that one has no energy left to analyse it. Begle's second law about mathematics education (1971, p. 30) applies especially to NLSMA:

Mathematics education is much more complicated than you expected even though you expected it to be more complicated than you expected.

Exercises

1 In reviewing NLSMA and similar curriculum comparison studies, Walker & Schaffarzick (1974, p. 97) observed that 'what these studies show, apparently, is *not* that the new curricula are uniformly superior to the old ones, though this may be true, but rather *that different curricula are associated with different patterns of achievement.* . . . Although this conclusion may seem obvious, a great many seemingly obvious generalizations about education have proven embarrassingly difficult to confirm by research.'

What are some examples of obvious generalisations about mathematics achievement that have not been supported by research? What are some plausible reasons why changes in the mathematics curriculum might have a stronger effect on pupils' achievement than changes in other features of instruction?

2 Noting the regularity with which the conclusion in the quotation above is overlooked, Walker & Schaffarzick ask (1974, p. 98): 'Why is it virtually impossible to find research which attempts to discover the consequences of studying different items of content, when there is so much research on the consequences of different media, methods, or strategies of teaching?'

What is your answer to their question?

3 Compare and contrast the taxonomies given in Wilson (1971). What are their strengths and weaknesses?

3.2. *Other studies of new-math projects*

Not all studies of New-Math curriculum development projects have been cast in the 'let's compare pupils' achievement' mould. SMSG itself was the subject of another kind of evaluation, the 'official biography' (see Wooton, 1965), as was the School Mathematics Project (see Thwaites, 1972). Less descriptive and more critical views of New-Math projects have been offered by DeMott (1964), Kline (1973), and the National Advisory Committee on Mathematical Education (CBMS, 1975), among others. These evaluations, both favourable and unfavourable, have tended to stress the justifications for undertaking a project more than the effects of its work. Although not considered by some researchers to be evaluations, these official biographies and unofficial critiques have influenced professional opinion.

In his classic paper on evaluation methodology, Scriven (1967) contrasted 'summative' evaluation, which looks at the final products of a curriculum development project, and 'formative' evaluation, which attempts to provide direction to the project's work. When this distinction was pointed out, most New-Math projects realised that they had been engaged in formative evaluation, although usually in a highly informal manner, when they tried out their ideas and materials in the classroom and asked teachers to report back on their successes and failures. In the Secondary School Mathematics Curriculum Improvement Study (SSMCIS), one of the case studies in chapter 3, the most important part of the evaluation conducted by the project itself was the monitoring of the materials' effectiveness by the teachers of the pilot classes. The teachers provided important feedback on how the materials were working, the problems they and their pupils were having, and so on. Achievement tests constructed to be used with the courses provided some data about concepts that were difficult for pupils to grasp, but the teachers' comments tended to be given more attention by the course developers than were the test results. Both SMSG and SSMCIS attempted at various times to provide some systematic information to writers who were revising a chapter as to how teachers thought the chapter should be changed, what pupils thought of it, and how pupils performed on test items covering the content of the chapter. Informal observation of writers' reactions to this information suggests that unless the information confirmed a writer's opinion of the chapter, he tended to disregard it in favour of his own private vision of how the chapter might be changed. In the face of evidence that teachers and pupils were having trouble with a chapter, writers tended to attribute the fault more to the teachers than to inherent defects in the materials. Several evaluators who attempted systematic and elaborate formative evaluations of New-Math materials emerged somewhat shaken from the experience of having their carefully prepared message disregarded.

The evaluation of the SSMCIS curriculum had one component that should be noted. SSMCIS was designed for the most capable secondary-school students, most of whom were bound for college or university. This meant that they would probably be required to take the College Board's Scholastic Aptitude Test (SAT) for college admission. Although the mathematics part of the SAT is not meant to be tied to particular school courses, the SSMCIS curriculum was so unusual that parents, teachers, and students soon became concerned that students who had followed it might be handicapped in taking the SAT.

Furthermore, SSMCIS students in New York State were required to take a special examination in order to obtain Regents Credit for their mathematics courses. Studies of the possible 'bias' of the SAT and the construction of special Regents examinations were a large part of the SSMCIS evaluation effort (Fehr, 1974) and took both time and resources away from other activities that might have contributed more directly to the quality of the curriculum. Some evaluation activities, however, are necessarily aimed at making a curriculum acceptable rather than at improving its quality.

Exercises

1 Since the emphasis of the New-Math projects was on improving the mathematical content of the curriculum, one can argue that some evaluation effort should have gone into analysing the content and getting experts to judge whether it had been improved.

   Do you know of any project in which this was done? Choose one of the New-Math projects and design a content evaluation for it.

2 In some countries, such as Federal Germany, the 'New Math' became a political issue, and candidates for office promised to do away with the new syllabus if they were elected. How does public opinion about the mathematics curriculum in your country get translated into action? What should be the public's role in curriculum evaluation?

3 In what sense can the booklet *Manipulative Skills in School Mathematics* published by the SMP in 1974 be seen as an exercise in evaluation?

4 Investigate the manner in which the 'chapter-revision' process has operated in other projects. What methods have been used to obtain feedback and how effective have these proved? Even if material has gone through two drafts (*cf.* SMP), is this any guarantee that material appearing in the published texts has been field-tested?

4.   The evaluation of projects based on the behaviourist approach

Mention has already been made in section 3, and an example given, of the use of various versions of Bloom's taxonomy to identify 'levels of cognitive process' that might be used in answering an examination question or a test item. Bloom's taxonomy was an attempt to develop a scheme for classifying both the outcomes of instruction (as measured by tests) and the objectives of instruction. The taxonomy is often taken as epitomising the behaviourist approach to evaluation: seek only those objectives for education that can be expressed behaviourally so that one can measure some aspect of the child's behaviour and tell whether he has attained the objective.

Numerous criticisms have been made of this approach, many of them by mathematics educators. Ormell (1974) has pointed out that although one should expect the attainment of educational objectives to be verified in terms of behaviour, one should not limit one's objectives to those that can be measured directly and reliably. Eisenberg (1975) argues more vehemently that behaviourism, by equating education with training, misses the essence of the discipline of mathematics.

Despite such complaints, and despite increasing expressions of dissatisfaction with 'the objectives model' in books and journals on educational evaluation (see, for example, Hamilton *et al.*, 1977), behaviourism still colours most evaluations of mathematics curricula.

Let us consider some evaluation studies of curriculum projects that espoused a behaviourist position. Not all of the evaluation studies themselves have been behaviourist in approach.

4.1. *The IPI evaluation programme*

The most comprehensive statement of the Individually Prescribed Instruction Project's evaluation activities is contained in a monograph by Lindvall & Cox.* The following excerpt (Lindvall & Cox, 1970, pp. 45–7) shows how the IPI procedure of devising a prescription (see pp. 141–50) for each student's work was given a 'formative' evaluation.

> *Instructional Prescriptions*
> An essential element in the IPI procedure is the procedure for developing, at frequent intervals, instructional plans or prescriptions tailored to the needs of the individual student. . . . The formative evaluation of the program gives major attention to the way in which these procedures are implemented and to the effectiveness of the prescriptions developed.
>
> (a) *Instructional prescriptions are based upon proper use of test results and specified prescription-writing procedures.* Evaluating this basic criterion has involved an examination of samples of prescriptions to determine if the procedures are being followed. Pretest results are compared with the objectives in which a pupil is assigned work to make sure that he has not already mastered those objectives.
>
> (b) *Instructional prescriptions provide learning experiences that are a challenge but permit regular progress.* Achieving this situation for each student is the specific goal a teacher seeks when he writes a

* C. M. Lindvall & R. C. Cox *The I.P.I. Evaluation Program*, 1970, Rand McNally and Co. Extracts reproduced by permission of Rand McNally and American Educational Research Association, Washington D.C.

prescription and is a major goal of the entire IPI procedure. . . .

A pupil's lack of progress may not be conclusive evidence that prescriptions are faulty since many factors contribute to progress or lack of progress. However, data can indicate cases where prescriptions and other elements should be reexamined. Conversely, evidence which shows that all pupils are making some progress does not necessarily indicate that prescriptions are satisfactory. . . .

(c) *Instructional prescriptions vary from pupil to pupil depending upon individual differences.* In terms of present knowledge of the instructional implications of individual differences, this quality can only be partially assessed. For example, information can be obtained to show whether or not prescriptions vary from pupil to pupil, a minimum criterion for evidence of this quality. Whether such variation is truly associated with individual differences must be decided after subjective analysis by the teachers and other staff. . . .

(d) *Instructional prescriptions permit pupils to proceed at an optimal rate.* The criterion which is represented here is central to the total IPI procedure. . . . Assessment efforts have been frustrating because of problems in identification of a reliable measure of rates and a definition of what is meant by 'optimal rate'. Early assumptions that this quality could be studied by examining correlations of rate with intelligence, reading ability, past achievement, and similar generally accepted correlates of current academic achievement were found to be faulty in view of an extensive investigation of relationships between these variables and various measures of rate showing that they were largely insignificant. Rate in IPI has been found to be essentially uncorrelated with traditional indices of academic aptitude. There is therefore a possibility that pupils are not proceeding at their optimal rates, or that rate measures are unreliable. True rate measures may not actually be a function of traditional measures of aptitude, or a pupil's rate may vary greatly from one learning task to another. Because of these difficulties, basic work continues on the identification of a reliable and meaningful measure of rate of learning. In the meantime, subjective analyses of the records of individual students provide information on the extent of successful tailoring of prescription to pupil rate.

(e) *Instructional prescriptions are interpreted and used correctly by the pupil.* This quality is investigated primarily through informal reports from teachers, teacher aides, and students themselves concerning problems that pupils encounter in trying to do what is prescribed. Informal feedback has been useful in revising prescription forms and in clarifying suggestions·for teacher use of the forms.

Summative evaluation ordinarily receives more attention than formative in evaluation reports because the latter tends to be informal and is intended primarily for the developers themselves. The summative evaluation activities of the IPI Project were extensive, as indicated in this excerpt from Lindvall & Cox (1970, pp. 59–60).

The assessment of the achievement of project goals

For the IPI program, evaluation of goal achievement has centered on the assessment of the six program goals. These goals are as follows:

    I. Every pupil makes regular progress towards mastery of instructional content.

    II. Every pupil proceeds to mastery of instructional content at an optimal rate.

    III. Every pupil is engaged in the learning process through active involvement.

    IV. The pupil is involved in learning activities that are wholly or partially self-directed and self-selected.

    V. The pupil plays a major role in evaluating the quality, extent, and rapidity of his progress toward mastery of successive areas of the learning continuum.

    VI. Different pupils work with different learning materials and techniques of instruction adapted to individual needs and learning styles.

It will be recognized that these objectives are what Scriven (1967) describes as 'intrinsic goals' rather than 'payoff goals'. These are goals that tell us something about how we can determine whether or not the programme is functioning successfully after the process of development and implementation have been completed. They are not goals stated in terms of what the students will be able to do after they have spent a given period of time studying under this type of instruction. This latter type of goal is found largely, in the IPI plan, in the extensive and detailed listing of instructional objectives. . . .

The six listed program goals are those that have given direction to the IPI program in its planning, development, and implementation and hence are the immediate goals that must be assessed in any final evaluation of the effectiveness of the IPI Project.

The IPI evaluation programme used various sources of information. The primary source was students' performance on the IPI tests – placement tests, pretests, 'curriculum embedded' tests, and post-tests. Standardised tests were also used, as were attitude inventories, classroom observations, and interviews with students, teachers, and parents.

According to Devaney & Thorn (1974, p. 121), independent evaluation studies of the IPI programme 'that involved comparisons on standardised tests in general indicated no significant differences between IPI and non-IPI students.' The results of studies of students' self-concepts after participating in the IPI programme were mixed.

Critical analyses of the IPI programme have been made by several observers, most of whom have simply visited demonstration schools in which the programme was under way. A common reaction is that expressed by Oettinger (1969, pp. 147–8).

> There is no denying the observation that children at Oakleaf and McAnnulty, another IPI school in Pittsburgh, were well disciplined though freely moving about, happy, and eager. On the other hand, they could frequently be observed marking time with a hand or a flag raised, waiting for the teacher to answer a question or for their turn to have tests scored by teacher aides, just as children mark time in ordinary schools.
>
> The children are happy and eager, but what are they learning? They are learning about that valuable but restricted range of human knowledge and attitudes that can be mechanically expressed and measured. Without other forms of education, they may grow up under the dangerous illusion that there always exists a correct answer to every question.

Probably the most devastating evaluation of IPI mathematics was done by Erlwanger (1973, 1974), who conducted extensive interviews with nine children in grades 4 to 6 using the scheme. The study began when, in the course of a visit to a grade 6 IPI class to assist pupils who were having trouble and to diagnose their difficulties, Erlwanger encountered Benny, a twelve-year-old boy whose teacher thought he was one of her best pupils in mathematics, but who turned out to have developed a number of misconceptions about the subject. For example, Benny had a variety of idiosyncratic rules for getting answers to problems involving computations with fractions or decimals. Sometimes the rules worked, but more often they did not, in which case Benny would formulate another rule that would yield the desired answer. Benny had come to see the purpose of school mathematics as getting the answer given in the answer book, even though that answer might be arbitrary and even unreasonable. The IPI programme, by cycling Benny through pages of repetitious exercises and tests, encouraged the strategy of trying a rule, finding out what the answer was supposed to be, and formulating another *ad hoc* rule that would yield the answer. By dint of diligent application of these rules, Benny was able to make better-than-average 'progress' without his teacher finding out that he did not understand the mathematics. What Benny did understand was that he had found a way to make rapid progress in a not-very-sensible enterprise where the answers were arbitrary and what counted was getting the 'official' answer. Erlwanger concluded that various factors in the IPI scheme had led to Benny's misconceptions, and he set out, through observations of and interviews with additional children, to see how prevalent such misconceptions were and what factors seemed to be responsible.

The case studies and Erlwanger's analysis of the IPI programme itself led him to make the following suggestions (Erlwanger, 1974, pp. 293–4).

> 'Its behavioristic approach to mathematics, its mode of instruction, its form of individualization, and its evaluation and diagnostic program inhibited the development of the children's intuitive ideas, [and] thereby encouraged the development of their misconceptions. The innovative features of the program appeared to produce effects in the classrooms such as: the elimination of traditional classroom practices like teacher demonstrations and group discussions; a change from a group social structure to an emphasis on individual progress; a shift in responsibility for learning from the teacher to the program and the child; conflicting roles for the teacher and the child; and so on. The case studies indicate that the prolonged exposure of the children to these changes may have encouraged the development of some of their ideas, beliefs and views. The teachers also appeared to be unaware of the effects of some of these changes upon the children. The case studies suggest that the dynamics of some of these interactions were too subtle to be detected through conventional procedures in evaluation and diagnosis, or through structured observation techniques. It seems that some of these interactions could only be identified and explained to the extent that the conceptions of children were understood.'

4.2. *Other studies of behaviourist projects*

The IMU Project discussed in chapter 3 can be classified as behaviourist. It was the subject of an extensive study conducted from 1968 to 1971 and reported by Larsson (1973). This research programme was unusual in that a complete plan for investigating its effects was published in advance (Jivén & Öreberg, 1968). A long list of variables was proposed as a basis for comparing three models: combined classes using IMU, single classes using IMU, and single classes using traditional materials. The variables referred to pupils' knowledge and proficiency in mathematics, study techniques, independence, and cooperation; pupils', teachers', assistants', and headteachers' experience of course content, material, methods, organisation, and the school situation; costs and other issues of implementation; the nature and quality of the instructional material, including pupils' and parents' views; the nature and quality of the instructional activities; and the performance of pupils with special needs.

The evaluators ran into problems in trying to compare the IMU with the traditional classes, for the materials of the former were geared to a new, revised curriculum, and those of the latter to a curriculum about to be replaced. This change also meant that teachers and assistants had to adapt not only to a new instructional system but also to changed course content. Moreover, the organisational patterns changed so much while the study was under way that their long-term effects could not be examined. A final problem was that the teachers participating in IMU had, for the most part, volunteered to adopt the scheme, and did not appear to be representative of the population of Swedish mathematics teachers in grades 7 to 9.

Nonetheless, with the aid of a variety of tests, questionnaires, observations, interviews, and school records some interesting findings were compiled about the project's effects. It was found that the degree of individualisation was high when it came to pupils' choice of booklets, the amount of individual instruction from the teacher, and the rate at which pupils worked; it was not so high with respect to variation in the difficulty of material, and it was quite low with respect to evaluation and marking. The pupils were generally favourable or neutral in attitude toward the IMU method, but as noted in chapter 3, both they and the teachers expressed dissatisfaction with the amount of group teaching. The teachers enjoyed being able to move around the class and give help to individual pupils, but they worried that pupils were not developing the ability to listen to and talk mathematics. They were also concerned that the formal treatment of mathematics had deteriorated under the IMU system. The success of the project in individualising the rate at which pupils worked created some administrative problems for the schools. Because of increasing numbers of pupils who lagged behind the others and seemed unable to work independently, some schools had introduced a minimum rate of work. On the other hand, some pupils progressed so rapidly through the programme that they created problems for the upper secondary school in giving them credit for their advanced work and in allowing them to continue from the point where they left off at the end of grade 9.

When comparing the 'combined' with the single class model, the evaluators found that the effects on the pupils were about the same regardless of class size, the numbers of teachers per class, or the number of rooms into which a class was divided. Teachers of single classes tended to spend more time on administration, whereas the other teachers tended to spend more time conferring and in group teaching.

The IMU evaluators had planned to analyse the mistakes pupils made in their booklets as a means of understanding their difficulties. The pupils, however, had an answer key, and the large number of cases in which errors in the key were repeated in the pupil's work convinced the evaluators that such an analysis would not yield valid results. This finding, regarding copying, although not emphasised by the evaluators, is a disturbing echo of Erlwanger's findings about IPI.

The main part of the IMU evaluation plan can be labelled behaviourist because of its emphasis on the effects of the IMU programme in the light of its objectives. But as the evaluation proceeded, circumstances forced the abandonment of any attempt to compare IMU teaching with conventional teaching. The most interesting results of the evaluation study concerned changes in the teacher's role in the IMU system and descriptions, from various sources, of the IMU material.

Computer Assisted Instruction drill-and-practice programmes in arithmetic are yet another manifestation of the behaviourist approach. Evaluations of CAI programmes have tended to emphasise gains in academic achievement in which experimental and control groups are given standardised tests as pre- and post-tests in some approximation to an experimental design. A contrasting kind of CAI evaluation study was conducted by Smith & Pohland (1974), who used an anthropological approach, and went as participant observers to five schools in the rural highlands of the Appalachian mountains during the 1968–9 school year. Their major finding was that the CAI programme seldom ran smoothly. Problems with malfunctioning teletypes and with the computers greatly limited the number of children who could have a drill lesson without disruption. Partly as a consequence, the teachers exhibited great variability in how they used CAI.

It was found that CAI had considerable social significance for the pupils, especially in the way various forms of competition developed.

Here is an excerpt from Smith & Pohland's field notes (1974, p. 31):

> Midway through the morning I happened to notice three boys working
> on the terminals. They made an effort to start together, and it was a
> real contest. It should be noted that the three boys were not on the
> same lesson. Nevertheless, there was a great deal of competition to see
> (1) who would finish first, and (2) who would get the highest
> percentage. The boy who finally did finish first raised his arms above
> his head like a boxer's and crowed rather exhaltedly, 'I won, I won'.
> The sweet smell of success was even greater when he found out that he
> had achieved a higher percentage score on his test than either of his
> two buddies. Both of them looked a little bit crestfallen, particularly
> the boy who ended up last.

Behaviourism as applied to curriculum development over the past two
decades has tended to emphasise the individualisation of instruction.
Just as the learning process can be broken down into small bits, so the
class can be resolved into a collection of individual learners. The social
consequences of such a resolution cannot be detected by means of
achievement tests and may not be picked up by attitude inventories or
questionnaires. Observations and interview studies of children undergoing individualised instruction have been powerful arguments against
limiting curriculum evaluations to paper-and-pencil instruments.

Exercises

1 In discussing the IPI criterion that pupils should be permitted to
   proceed through the programme at an optimal rate, Lindvall & Cox
   (1970, p. 47) observe: 'There is . . . a possibility that pupils are not
   proceeding at their optimal rates, or that rate measures are unreliable.
   True rate measures may not actually be a function of traditional
   measures of aptitude, or a pupil's rate may vary greatly from one
   learning task to another. Because of these difficulties, basic work
   continues in the identification of a reliable and meaningful measure of
   rate of learning.'
   What is your (the reader's) true rate of learning mathematics? What
   are the prospects for identifying a reliable and meaningful measure of
   rate of learning mathematics?

2 The problems encountered by those IMU pupils who worked rapidly
   and then did not fit into the upper secondary mathematics programme
   are reminiscent of the problems found by the SSMCIS pupils who
   entered college already familiar with much of the first two years of
   undergraduate mathematics. Outline a study to evaluate the effects on

pupils of transferring from an accelerated mathematics curriculum to a more traditional curriculum.

3 'Evaluators who are staff members of mathematics curriculum development projects based on the behaviourist approach have used the "agriculture–botany paradigm" in evaluating the project's work, whereas outside evaluators have used the "social–anthropological paradigm".' Locate two evaluation studies that support the preceding generalisation, and then try to locate two evaluation studies that refute it.

4 Joseph Lipson, who supervised the development of the first versions of the IPI mathematics modules, later concluded that the programme was built on false assumptions that ignored some of the advantages of group instruction. In particular, he noted that with IPI, students do not get a sense of the total mathematics curriculum. In his words, 'by focusing on the IPI module, the student hears the individual notes but fails to detect the melody' (Lipson, 1974, p. 61). What are some methods teachers can use to help students detect the melody? Can these methods be incorporated into modular schemes?

5 Is it only the behaviourist approach which implants in children the belief that 'there always exists a correct answer to every question'? How can this danger be avoided?

6 Bloom concerned himself with the 'expected' returns of education. Is it possible to evaluate 'unexpected' outcomes?

7 Write an essay on 'The ability to listen to and talk mathematics'.

5.   **The evaluation of projects based on the structuralist approach**

Mathematical curriculum development projects that could be labelled 'structuralist' tended to share with the 'New-Math' projects a mistrust of evaluation and a belief that although informal evaluation activities might be undertaken to help improve the project's work, formal evaluation activities were likely to be misleading. Unlike curriculum developers of the 'behaviourist' persuasion, who welcomed evaluation as an integral part of their work, 'structuralist' and 'New-Math' curriculum developers tended to look askance at evaluation studies, especially those undertaken by outsiders. They submitted to evaluation as a necessary price for continued support by a funding agency, but they worried that the subtle kinds of 'higher-order' learning sought by their projects would not be easily detected. They saw their project's work as virtually self-validating. The materials they developed were intrinsically worthwhile and seemed to be usable. Why was a 'full-dress' evaluation necessary except to satisfy the bureaucrats?

The evaluation studies of 'learning by discovery' seemed to confirm their scepticism. Empirical studies comparing learning by discovery and learning by exposition have yielded conflicting findings attributable not merely to weaknesses in the design of the studies themselves but also to problems of defining the two approaches and deciding on criteria for comparison (see Shulman & Keislar, 1966, for discussions of these problems). The difficulty of comparing something as amorphous as 'the discovery method of teaching' with any other teaching method seemed to convince curriculum developers who advocated a discovery approach that they should not permit their projects to stand or fall on the basis of an experimental comparison of teaching methods.

Projects based on the structuralist approach have, therefore, seldom received extensive formal evaluation. An exception is the Comprehensive School Mathematics Project (CSMP).

5.1. *The CSMP Evaluation Program*

In the autumn of 1972, all programmes of the research and development centres and regional educational laboratories of the USA were reviewed as part of their transition from the Office of Education to the National Institute of Education. The ten-member panel that reviewed CSMP recommended that funding for the project be terminated from June 1973. The director of CSMP, Burt Kaufman, and his collaborators and colleagues around the world quickly raised a great outcry, arguing that the work of CSMP had been unfairly judged and accusing the panel of making a behaviourist attack on humanistic mathematics. (The November 1973 issue of *Educational Technology*, entitled 'A Response to Managerial Education', is devoted to articles setting forth this theme.)

The National Institute of Education, responding to this argument and pressure, reduced the funding rather than terminating it and decreed that CSMP's large-scale curriculum development activities would be suspended pending the results of a summative evaluation programme to begin in 1973.

The CSMP Evaluation Program was directed by Martin Herbert with the assistance of a five-member evaluation panel. The specific issues to which the evaluation was directed were as follows (Herbert, 1974*a*, p. 33):

1. *Intrinsic Merit*
What is the opinion of qualified reviewers regarding the soundness and relevance of the program and its mathematical content?

2. *Practicality*

(a) What is the cost, to adopting school systems, of buying materials (one time cost) and of maintaining materials for continued usage? How does this compare with present materials cost for elementary school mathematics instruction?

(b) What are the personnel requirements for a school system in using the program? (This applies particularly to the role and duties of the local coordinator, to the time and effort required for teacher training and to the possible need and availability of teachers with more specialized training in third grade and beyond.)

(c) How successful are teachers in coping with the program and in implementing it in a manner reasonably faithful to the intentions of CSMP?

(d) Do users like the program in comparison to other mathematics programs they have used?

(e) Can students transfer into and out of the program at any point in the curriculum without creating serious difficulties for the student or his new teacher?

3. *Outcomes*

(a) Do students learn the basic concepts and skills, particularly computational skills, generally expected of students in elementary school?

(b) Do students learn the specific skills and concepts of the CSMP curriculum or, otherwise stated, do they attain the behavioral objectives of the program?

(c) Are CSMP students, particularly after two or three years in the program, better able to deal with certain kinds of mathematical situations than are students who have not studied CSMP?

The issue of intrinsic merit was addressed by asking five members of the Mathematical Association of America to examine the CSMP materials and evaluate the soundness and relevance of their mathematical content. Shirley Hill was the chairman of the group and summarised the reviews they wrote. An excerpt from her summary gives something of the flavour of the external review (Herbert, 1974*b*, pp. 3–4):

> The overall impression of the materials was favorable; three reviewers expressed quite favorable evaluations directly, the reaction of another was mixed, and the impression of the fifth cannot be said to be favorable, though it was not explicitly negative.
>
> One point of general agreement in the reports was on the soundness of the mathematical content. The material is seen to be mathematically sound without any egregious technical or conceptual errors. There

were differences of opinion concerning matters of preference and taste in the development of the mathematical ideas.

It was at least implicit in every report that it was impossible to separate completely in an evaluation of this kind, matters of mathematics and matters of pedagogy. Certainly most of the differences in preference concerning the way the mathematics was presented had little to do with mathematical soundness but rather related to questions of learning, development, concept formation and the like. Many of these are empirical questions. I think that it is fair to say that *most* of the very specific comments and specific criticisms concern psychological and pedagogical issues.

An example of a curricular element which is a mix of mathematical and pedagogical issues is the use of the minicomputer [see Papy, 1969]. This is the single point of complete agreement among all reports. There is too much reliance on the minicomputer. Three reviewers vehemently opposed its use as an aid altogether; the other two seriously question its value in light of the very great investment of time. (Both of these reviewers agree that the effectiveness of the device with respect to computational skills is an empirical question.) All five reviewers are dubious to *very* negative on the minicomputer's mixture of a binary and decimal base.

Are the materials innovative, current, timely? Comments ranged from 'it is more of the same' to 'the material is refreshingly full of new ideas'. The majority were of the opinion that the materials were timely and current and in many instances excitingly new. One reviewer found much new material of which he could approve but too much 'old' material from the era of 'new math'. One found some 'good sections' but little mathematics and much 'obsessive ritual'.

The remainder of the evaluation is not easily summarised. The CSMP programme seemed to be well liked by the teachers in the Pilot Trial schools. Many teachers, however, thought it too difficult for their low ability pupils, and they tended to proceed at a slower pace than was recommended. Classroom observers indicated that, except for some problems of pace, the programme was implemented as intended. Reviewers judged it to be reasonably practical. The per pupil cost of the CSMP kindergarten programme was relatively low, but the CSMP first and second grade programmes were more expensive than most of the others surveyed. Both the quality of the teacher training and the role of the local coordinator were judged highly variable in the Pilot Trial schools. The pupils' learning was generally adequate, and was excellent in some cases. Pupils in CSMP classes usually outperformed pupils in comparison classes on achievement tests of various kinds.

Because the interviews with teachers and the observations of their classes are a unique feature of the CSMP evaluation, we reprint below some excerpts. First, we reprint excerpts* from an interview conducted at the end of the school year with a second grade teacher (Barszcz, pp. 89–90). The remaining excerpt is a 'brief sketch' of a first grade class based on observations, interviews with pupils and teachers, and field notes (Holz, Herbert & Karmos, 1974, pp. 61–2).

I: Now that you have taught the CSMP program for nearly a year what is your general impression of it?

T: I think it is very good and I think that the children who are average and above average thoroughly enjoy it and they have gotten much from it. But I really feel that the children who are at the lower end are very much confused at times and aren't really picking up any of the things that we are doing. Now I don't know if those children would be confused with another program or not. But some of the things that they were to have learned in the first grade they didn't pick up there which made it harder in the second grade program for them to move on. Now those children haven't had as much trouble as children who have come in the middle of the year. Those children had a great deal of trouble picking up this whole concept. And the Minicomputer was almost impossible for them and I didn't even attempt to teach it after a while when I saw that they weren't getting it. Now these are slow children and there again that could very well be the reason for the difficulty.

I: Now, you said that the average children and above average children really like it. What aspects of the program do you think that they liked the most? What turned them on to the program?

T: That's something I'd like to know myself. I don't know really. At the very beginning of the school year I said it's time for math [and] they were ready to go. I don't know what it is really. They liked the Minicomputer, to them that was fascinating.

I: In what ways did your perceptions of the program change throughout the year since the beginning when you first started teaching the program? Has it changed at all?

T: Well I guess I kind of see it more as the whole thing. Perhaps at the beginning I should have gone through and looked through it but that was a tremendous task which I didn't have time for. But I kind of see the whole thing now as a whole, rather than step by step, lesson by lesson which should make it much better for me next year.

. . .

I: What would be some of the worst aspects of the CSMP program?

T: Well the thing that I really didn't like and maybe I didn't like it with the

* Reproduced by permission of CEMREL Inc. St. Louis, Missouri, from CEMREL'S Comprehensive School Mathematics Program.

slow children was the spiral approach. A lesson is introduced and then not touched again for a week. I think the slow child really needs more repetition at the time it is introduced. This is what I found. And then I would have to take a slow group and really work on that idea or it would be totally foreign the next time it was introduced to them. I would say maybe a little more on that one thing before you leave it and move on to something else for some children. Now it didn't present a problem for the faster children at all. . . .

Class 14
This class was actually two normal sized first grade classes combined, utilized two classrooms, and was team taught by two teachers. One of the classrooms contained all of the students' desks and chairs. This was where formal sit-down classes took place. The other room which was nearly empty was used for physical activities, art, etc.

There were over 50 children in the class. The average rank was at the 40th percentile on the pretest. The children seemed extremely 'at home' and relaxed in the class. They were very open with each other, the teachers, and the observers. The class had an atmosphere of creative chaos which was refreshing and interesting to observe.

One of the teachers was a young woman and the other a young man. Both were lively, open, and friendly with the children. They had a direct manner of dealing with the children who obviously enjoyed their teachers.

During observed sessions one of the teachers would 'teach' the lesson to the class while the other teacher would move around the room helping individual children, prodding some children to pay attention, and often interjecting comments into the lesson. The lessons moved briskly and a surprising number of children were given the opportunity to respond. The children seemed enthusiastic and eager. The teachers appeared to try to get everyone in on the act. Usually the lessons followed the plan quite closely.

During work periods when children were doing workbooks and worksheets on their own both teachers would help the children and check work as it was being done. However, there were usually more questions than could be handled by the two teachers. The children were generally allowed to work together during these sessions and they did this remarkably well. The class was usually very noisy.

Both of the teachers were very pleased with the program. They thought their children learned much more than in previous classes. They were the only teachers who felt the program was especially good for slower children. In fact, they had two older children in special education classes in the school come to their class just for math. They claimed that for the first time these children were achieving something in math. They also found that one of their very slow children did so well on the placement test in math that although she is to be placed in a special class for other subjects she will remain in CSMP math next year. This had *never* happened before. They explained that the slow children did not understand everything and did not do as well as average and above children, but they did much better than children of equal ability had ever done before.

Despite the careful and generally favourable summative evaluation of CSMP, funds remained hard to obtain. The government panel's review of the CSMP materials in 1972 proved to be a more powerful evaluation than any set of findings based on test scores, teacher interviews, or classroom observations.

Exercises

1 The CSMP evaluation was directed at three major issues: intrinsic merit, practicality, and outcomes. Is this an improvement over the behaviourist evaluator's concern with the attainment of specified goals? Why or why not?

2 Examine the November 1973 issue of *Educational Technology* and the responses that appeared in subsequent issues. Can you devise a behaviourist's response that might have been given to the arguments put forward by Atkin and by House with respect to evaluation?

3 Obtain more information about the Papy minicomputer used in the CSMP programme. Suppose you were responsible for revising the CSMP's use of the minicomputer. What would you do about it in light of the reservations expressed about it by the reviewers; the finding that whereas many pupils learned to use it correctly and well, many others could do virtually nothing with it; and the finding of a clear dichotomy between teachers who said it worked well at all ability levels and those who said only the high-ability students could use it?

4 The teacher interviewed admitted that she had not read through the year's work before commencing to teach it. Discuss the professional teacher's responsibilities to a project whose materials she is using and to the children she is teaching. What is it reasonable for a project to ask and expect?

5 In Britain, where the 'spiral curriculum' has operated for many years, the time spent on a topic before 'moving on' has tended recently to decrease (*cf.* SMP Books 1–5 and Books A–H). Discuss the possible effects of this as it concerns high-ability, average and low-ability children. (Consider the remarks of the interviewee.)

6 Begle (1979, p. 128) dismisses 'team teaching' as having been clearly shown to have 'no advantages over conventional teaching'. Discuss the

implications of Begle's statement in the light of your experience and the case study reprinted above.

6.   The evaluation of projects based on the formative approach

The formative approach to curriculum development requires that teachers play a strong role in helping to devise situations in which children can learn mathematics. Projects based on the formative approach tended to provide materials that were illustrative and suggestive rather than comprehensive and definitive. They left much of the developmental work to the teacher, who was often expected to treat the materials as a supplement to the regular mathematics programme.

Such projects were difficult to evaluate. They encouraged freedom and diversity in implementation, so one class's curriculum might be quite different from another's. Because they seldom were concerned with all the topics in the syllabus, they often could not be considered a complete programme and judged accordingly. Such projects usually did not set forth a list of goals, but when they did, the goals tended to deal with vague, albeit important, notions about such things as learning to pursue mathematics on one's own and developing an educated intuition. The projects rarely set forth a detailed statement concerning how progress toward such goals might be assessed.

Thus, for example, the Madison Project did not attempt systematic evaluation studies of its own work. Robert Davis (1965, 1967) cited some modest research that compared the achievement test performance of Madison Project classes with that of a class of pupils who had not studied Madison Project materials, but such research clearly played a minor part in Davis's ideas about evaluating the project. As evidence that the Madison Project was successful in changing the curriculum, Davis listed the curriculum development programmes influenced by the project; cited workshops, institutes, and professional meetings in which members of the project participated; indicated that the project had frequently been mentioned in professional publications and by the news media; and noted growing numbers of participants at project-sponsored events. Davis described various methods (such as getting reports from teachers and checking its pedagogical rationale against popular models for human behaviour) by which the project attempted to evaluate the desirability of the changes it was attempting, but he did not spell out the results of the project's evaluation activities. After listing nine methods used by the project to determine the undesirable outcomes that might have resulted from the introduction of Madison Project materials (for example, tape recording of lessons and interviews with students and teachers), Davis (1965, pp. 97–8) drew the following conclusions:

> There is, at present, no evidence of undesirable outcomes beyond one or two which are inevitable.
>
> (i) Use of any new curriculum involves *decisions,* and any issue within a school which involves decisions creates the possibility of disagreement. The question of coping with decisions and disagreement is a continuing question for every school system, and must be dealt with in a satisfactory way if the school is to rise above mediocrity.
>
> (ii) Use of 'new mathematics curriculum' materials of any sort raises questions of cooperation among teachers, especially regarding articulation from one year to the next in the student's life. The question of faculty cooperation, also, must be resolved if a school is to achieve quality education.
>
> (iii) Although no cases have been observed, it seems safe to predict that if a school 'adopts' *any* new curriculum materials without adequate teacher education, without adequate administrative support, or without an adequate determination to cooperate and to succeed, the results may well be abortive, disillusioning, and detrimental to further curriculum efforts of any sort.
>
> (iv) There is always danger of 'nullification by partial assimilation', 'tokenism', or 'complacency'. A school may pick up a small piece of new curriculum, and then, with a sigh of utter accomplishment, cease any further efforts. Curriculum evolution must, somehow, become an on-going activity for every teacher and every school.

Unfortunately, Davis did not present any of the evidence that he used in deciding that the project had led to no undesirable outcomes beyond these. He did not report in any detail the results of applying the nine methods, and he did not even indicate the extent to which inevitable disagreements, lack of cooperation, or complacency were observed in schools using Madison Project methods and materials. In short, although Davis appeared to recognise the importance of an evaluation of his project, he did not attempt to probe beyond superficial indicators of success.

The evaluation of one project whose approach could be labelled 'formative' did, however, produce some interesting results. This was the Californian Specialized Teacher Program, part of the Miller Mathematics Improvement Program, which provided summer workshops to introduce elementary school teachers to new approaches to teaching mathematics. The workshops, like those of the Madison Project, emphasised activity, the use of apparatus, and discovery learning. Participants then became specialist mathematics teachers, teaching the subject not only to their own classes but also to those of colleagues. Several evaluation studies of the project were conducted by Dilworth & Warren (1973), who found that pupils whose teachers had participated in the workshops tended to score higher on various tests of computation and understanding than pupils whose teachers had volunteered for the workshops but could not be accommodated. Pupils of teachers who had attended workshops for two summers tended to do better than those of teachers who had attended for one summer only. These evaluation studies are notable for the care with which they were conducted and, in particular, for the random assignment of teachers to groups for training. They clearly demonstrated the way in which the training scheme for teachers yielded relatively consistent and reliable gains in pupils' achievement: yields which are frequently sought but which have only rarely been shown to have been achieved.

6.1. *An evaluation of Nuffield mathematics*

Like other projects based on the formative approach, the Nuffield Mathematics Teaching Project (5–13) received little systematic evaluation. A Schools Council (1974) evaluation of primary school mathematics found that after ten years the project had spawned a variety of practices, but no distinctive orthodoxy had emerged.

Hewton (1975) presented an evaluation based on sales indicators, interviews with people associated with the project, and statements in books, journals and the press. The most detailed analysis was that based on figures from 1967 to 1972 showing the sales of Nuffield materials in the United Kingdom and elsewhere. The figures showed that the sales per year for all countries peaked in 1968 and declined thereafter. Declining sales in the UK were somewhat offset by increasing export sales, especially in the USA. When the estimated number of annual adopters in the UK was expressed as a percent of the total market, the data resembled the S-curve commonly found in studies of the diffusion of innovations: a small number of early adoptions is followed by a surge and then a decline. In the case of Nuffield mathematics, however, the surge came almost immediately. Hewton speculated that activities during the trial period created a 'pent up' demand, that many factors made innovation in mathematics popular at that time, and that the lack of competition stimulated initial sales. The sales figures suggested that perhaps half of all primary teachers in the UK may have purchased Nuffield mathematics materials at some time. Estimates of the spread of Nuffield ideas into primary schools suggested that from 50 to 75% of the schools had been influenced in some way.

Hewton noted that over 600 teachers' centres had been set up as outgrowths of the mathematics centres sponsored by the project. He concluded that the teachers' centres had provided much valuable support and encouragement to teachers in the early days of the project but that they were not sufficient to maintain such support unaided. He also judged the quality of the in-service courses for teachers provided by the centres to be mixed. The project's introduction of a 'concept map' into the guides, and its publication of a 'guide to the guides', were responses to pressure for more detailed guidance in implementing the curriculum.

The Nuffield mathematics project was faced with continual pressure for more pupils' materials, but it resisted this pressure because the project wanted teachers to develop their own approach and materials. Hewton found that this demand was eventually met by materials produced outside the project, and that even some of the project's supporters charged it with insensitivity to teachers' needs in this respect.

Hewton dealt with the charge that Nuffield mathematics was responsible for pupils' poor performance in computation by asserting that performance had always been poor. He suggested that the project might have been unsuccessful in convincing teachers and parents that new attitudes were needed toward numeracy and computational skill. Nonetheless, the project appeared to have made an extensive impact on education, especially when its relatively modest research and development costs were considered.

Exercises

1 The developers of projects based on the formative approach have tended to dismiss evaluation studies that have examined the achievement of pupils in such projects, whereas 'outside' evaluations of the projects have tended to stress pupils' achievement as the criterion of success. Recall the suggestion in exercise 3 of section 4 that the opposite situation may hold for projects espousing a behaviourist approach. How safe is the generalisation that 'inside' evaluations reflect the approach of the curriculum developer, whereas 'outside' evaluations reflect a contrary approach?

2 Using data on the number of candidates for O- and A-level

examinations for Nuffield Science projects and the School Mathematics Project, Tebbutt (1978) offers several models for the diffusion of these projects' impact. One of the difficulties these models have is accounting for a shift from a rapid rate of exponential growth at the outset to a lower rate of exponential growth. Hewton attributed a similar phenomenon noted in sales of Nuffield mathematics materials to 'pent up' demand. What other factors might account for the rapid initial rate with which curriculum innovations are adopted? Tebbutt ignores those schools which adopted the SMP materials but made use of 'modern' O-level examinations other than that set by the Oxford and Cambridge Board for SMP. What happens to his model and conclusions if one takes such schools and their pupils into account?

Tebbutt draws attention to the different patterns of take-up of SMP O- and A-level materials, but makes no attempt to explain these on mathematical/educational grounds. What explanation could you offer?

3   'It soon became clear that we had to go beyond labels like 'Nuffield', 'Dienes' or traditional . . . now most schools employ – quite rightly – a mixture of methods' (Ward, 1979, p. 12). This elaboration of Schools Council (1974) gives its approval to the idea of a 'mix' (a safe British compromise). In what sense can one select bits and pieces from Dienes, or Nuffield? What mathematical and pedagogical expertise is required to make an informed choice? What features would you expect in an 'uninformed' mixture?

6.2. *Alternative modes of evaluation*

The Fife Mathematics Project, one of the projects studied in chapter 3, can be classed as being based on the formative approach because of its stress on teaching method rather than mathematical content. It was evaluated by Douglas Crawford (1975) and a later investigation by Judy Morgan (1977) also made reference to it. Crawford tried to make his evaluation 'illuminative' (Parlett & Hamilton, 1977). He obtained a description of the project from its organiser, and accounts of how it operated in the classroom from teachers in a representative set of project schools. Pupils' reactions were collected by teachers who themselves were asked to complete questionnaires concerning project work in the classroom – its organisation and its effect on teaching – and changes in the teacher's attitudes toward the project after a year's experience. A test of mathematical achievement was given to the pupils, but the results proved difficult to interpret since the amount of time pupils spent on project work varied greatly. Personal assessments of the project were made by teachers and headteachers in several schools and by Crawford himself.

Crawford's report allows the reader to form a good idea of what it was like to participate in the Fife Project, either as a teacher or as a pupil. What it does not offer are definite conclusions or detailed consideration of particular issues. The excerpts below,* taken from Crawford's summary (Crawford, 1975, pp. 129–33), give some indication of the conclusions which were drawn.

How successful is the Project?

Content, materials, and method
One of the most significant aspects of the Project is its *practical* approach to mathematics. Pupils learn by *doing*. According to Piaget, the Swiss psychologist, children go through several stages of mental development. He claims (and this is generally accepted) that concepts are formed as internalized actions. Children need the physical experience of handling concrete objects if they are to form appropriate mathematical concepts. Reforms at primary school level such as the Nuffield project have been based on Piaget's work, but there has not been a similar recognition and application of it at the early secondary stages, when many pupils have not yet reached what Piaget calls the stage of 'formal operations'. Only after the pupil has developed well-established concepts, which he can manipulate and apply without recourse to concrete experience, can he think entirely on an abstract level.

George Sampson underlines the value of the Project's practical approach. By giving children many concrete and visual experiences with shapes, areas, and solids, they can build up a much clearer idea of what is meant by angle, area, three-dimensional, and so on. Moreover, pupils seem to enjoy 'doing things', though the impatience of some high flyers suggests that they have already reached the stage of formal operations when a verbal and bookish approach is more appropriate.

When we look at the contributions from teachers we find that the comments regarding the quality and suitability of the mathematical content are generally favourable. . . . But there are several assertions that some of the material is too difficult, especially at the very early stages. Another criticism is that more explanation is needed and that the language is not always clear.

The pupils seem satisfied with the content on the whole. Most of them seem to find it neither too difficult, nor too easy. A number of them say that they enjoy handling the materials. But the top set at St. Columba's High School in Dunfermline say that the booklets are too easy and that there is too much emphasis on working with concrete

* Reprinted with permission of Oxford University Press, from The Fife Mathematics Project, Ed. D. H. Crawford, © Oxford University Press, 1975.

materials. Generally speaking, the average and less able pupils seem to be well satisfied, while some of the brighter ones are more doubtful. Most of the teachers seem to feel that pupils find the materials interesting and stimulating.

The evidence suggests that the booklets are pitched at about the right level for most pupils, and my own observation of the Project supports this conclusion. . . .

Turning now to the method, let us first agree on what it is. I shall take it to mean the way of working in the classroom which accompanies the use of the booklets and other materials. This makes a clear distinction between 'Project method' and 'class-teaching method'. According to the first Stirling questionnaire, most teachers seem to prefer a mixture of project and class teaching. . . .

It is clear from the many favourable comments from pupils that the vast majority do enjoy the Project method of learning mathematics. They appreciate the opportunity of choosing their own work and being able to work at it in their own time. Obviously the Project allows the pupil far more personal responsibility than he has in other lessons, and many of the comments suggest that the pupils are learning how to work and think on their own, which though difficult to measure quantitatively, will be very valuable in future.

The teacher: discipline, organisation, and assessment

. . . (It) should be clear to the reader that the way of learning and teaching during a Project period is radically different from a traditional lesson. No longer do all pupils work on the same topic or set of exercises. The job of supervising and guiding the numerous different tasks on which pupils are engaged is no easy one. The teacher's *successful control* of the classroom will depend to a large extent on how well he *organises* himself and his pupils, and also on how well he solves the problems of *assessment* when using the Project.

Most of the teachers' contributions refer to all three aspects. In general, they feel that discipline problems do not arise as long as the teacher is experienced. . . . But organisation is of prime importance. Materials must be readily accessible, seating arrangements must be flexible, and the procedure for using the booklets and other equipment must be clearly understood. . . .

The problem of marking and checking is also an important issue. Assessment presents considerable difficulty. . . . Some pupils are worried about queuing. To overcome this, teachers should make themselves fully familiar with all the booklets, and so spend only short periods of time with each pupil. It is also important that they should circulate in the classroom, making contact with each pupil and anticipating problems before they arise. It is their responsibility to organize marking, especially of post-tests, so that they can supervise the classroom adequately. . . .

The pupils: achievement, attitudes, and development
Perhaps the most important question is 'How well does the pupil learn when using the Project?' Even here we run into problems; what is he learning? He may learn to pass examinations, but not how to work on his own. Which is more desirable?

The Project began as an open-ended experiment. It was not intended to be related to the S1 syllabus, though obviously certain ideas and concepts were likely to be encountered again later. Despite pressure to make the content more closely linked with the S1 syllabus, this principle has so far been retained, and has strong support from teachers who have been associated with the Project from the beginning. According to the 1971 Woodmill experiment and, more recently, to individual teachers and the Edinburgh University assessment, pupils using the Project do not appear to achieve less with the S1 syllabus than non-Project pupils. So far the traditional S1 syllabus has been the only criterion; the Project work has not yet been fairly assessed on its own ground. . . . Stirling questionnaires and comments from pupils and teachers indicate that most children have a more favourable attitude to mathematics in general. As already mentioned this teaching method seems to encourage more personal responsibility.

Summary
On the basis of these criteria, the Project seems to have been fairly successful in solving some of the more immediate problems associated with mixed-ability teaching, and in doing so without detriment to the pupil's expected achievement in the traditional setting. Whether the wider educational aims . . . have been achieved has not been reliably demonstrated, although there is much favourable subjective evidence. The Project clearly makes new demands on the teacher, but there is no sign that this is causing a significant problem. The teachers have risen to the challenge it involves, or at least have learnt to cope with the difficulties.

The enquiry carried out by Morgan under the sponsorship of the Scottish Education Department was primarily concerned with the affective consequences of the SED individualised learning project. Although this project can hardly be classified as formative, it would seem advantageous to discuss Morgan's enquiry here since it also concerns the Fife Mathematics Project and it provides another example of illuminative evaluation.

Rather than plunging immediately into traditional research activities by which data are gathered according to some plan, Morgan began by considering, from her viewpoint as an experienced teacher, the problems posed when the label 'individualised learning' is given to a method of learning. School visits and videotaped observations convinced her that the label meant different things in the Fife and SED projects. She decided, therefore, to look at only one of the projects 'because I knew that "individualised learning" was not a single teaching method and that there was no set of characteristics which distinguished it exactly and precisely from any other method of teaching mathematics' (Morgan, 1977, p. 7).

Morgan's observations of the SED Pilot Project led her to attempt an analysis of the conflicts felt by teachers regarding changes in teaching methods and the children's reactions to changes in the relationship between teachers and pupils. Morgan offers no neat summary or conclusion. The value of her work is in the depth of her analysis. We quote from her discussion of conflicts that the use of 'self-explanatory' work cards could cause for a teacher who preferred streamed classes and who was more concerned with the content to be learned than with the learner (Morgan, 1977, pp. 44–5).

(i) **The notion of work cards for children being self-explanatory**
When the Teacher is told at meetings that the cards are intended to be self-explanatory, when he reads the material and finds written down all that he knows he usually says, he finds it easy to believe that the cards and sheets are in fact self-explanatory. This is generally taken to mean that, providing the child can read, he can understand the work he has to do. If the child does not understand it, it will be because of his lack of intelligence. . . . The Teacher forgets the intrinsic ambiguity of language, the personal nature of interpretation and the need for the newly observed images to form themselves into a pattern which has some fit with previous patterns. All of these are idiosyncratic to the individual.

When the same Teacher teaches a class from the blackboard, he not only *says the words*, he points, sighs, raises his eyebrows, laughs, does all manner of things of which he is not conscious but *all of these* are part of the pattern of the new topic forming in the child's mind. When, later on, the Teacher wants to refer to the topic he repeats some of the emphasis, some of the gestures and these may be the very things which allow the child to retrieve the previous pattern, what we usually call remember (remember – reform, form again, organise members again). The child will not have the work, himself and the teacher in dissociated packs. He is affected by the whole being of the teacher-explaining-the-work-to-him. When he reads the sheets and cards, which must inevitably assume facts and skills hitherto experienced, only some of the new explanations form themselves into recognisable patterns which fit with patterns from his past. There are fewer characteristics available for ordering and playing with in his mind; there are only written symbols. . . . Now let us turn back to the Teacher. He believes that the cards are self-explanatory, he has been told that this is intended to be the case and he recognises the words used since they are very similar to the words he uses and has heard himself in connection with that particular topic. Because of all this he has no real cause to question the contention that the cards are self explanatory.

When he feels he is no longer needed as a person-who-explains-the-topic, he becomes uncertain about what to do as a teacher. His function seems to be diminished. . . . For this Teacher, a teacher who has a work-centred thinking space, an ability to explain a topic is not merely important, it is the major part of his teaching function. When he thinks of himself teaching, he thinks of himself explaining-the-work-to-the-*class*. He does not think of himself as a person who might make connections, associations between the work and a particular child. Nor as a person who *is the connection* between the work and a particular child. . . .

It is perfectly reasonable for the Teacher looking at the Project in this way, to ask what his role is. He sees himself as the person who puts out the cards, sheets, apparatus; marks the exercises, corrects the tests, fills in the assignment forms, maintains classroom discipline. 'I am no longer a teacher', he says. 'I miss hearing the sound of my own voice'; 'I miss teaching from the blackboard'. The challenge he finds in teaching is analysing each particular topic into simple logical steps which lead inevitably to the conclusions he has in mind; and can be spoken in good, easy-to-follow-explanations; and maybe also in entertaining thirty children in separate time packages of 35 minutes or an hour or whatever. . . . A teacher who believes that he is no longer a teacher is no longer committed to any of the children in any way.

The next excerpt is taken from Morgan's discussion (pp. 48–56) of how the format of work sheets and work cards can cause conflict for the teacher regarding his role, and how it can push the children toward getting answers fast, at the expense of understanding.

The format of the sheets and cards

The emphasis in the design of the sheets is on children 'doing mathematics' and not having to waste time copying out vast quantities of seemingly irrelevant stuff. This is definitely appreciated by the children who very often mention the fact that 'you don't have to write a lot' as a point in favour of the sheets, and a point of preference over the cards. The implementation of the ideal of 'less waste of time' is achieved by having the children fill in blank spaces rather than copy out whole tables or whole examples.

. . . (Only) very rarely does a child have to write out a complete answer to a problem and even then there is often a pattern of an example to copy. . . . Now I want to consider possible reactions of the Teacher to this format before considering the children's reactions in more detail. The first important difference is that in all his previous teaching life and in all lessons other than the Project lessons, the Teacher spends a lot of time making children write out their answers to the problems correctly, with a very definite format in mind. He is making them learn to write mathematics as it should be written, as he has always seen it written. . . . He is preparing the children for an examination where definite limits are set upon the presentation of answers to problems and this is his very reasonable justification, if he requires one. But it is not immaterial that he may think *all* mathematics is written in that particular form which he rehearses with the children.

Now, with the Project modules he is required to accept answers only, or shortened forms of answers. A child has traditionally demonstrated his understanding of a problem and its answer by his ability to write out the answer in a required mathematical form showing several stages of his reasoning. This activity is sometimes allowed for in the sheets, especially with more complicated problems, because certain key steps are left as blanks to fill in. However, there are not as many of these examples as would normally be done by a child using text-book resources. The Teacher no longer feels that he knows that the child understands what he is doing. The Teacher begins to ask questions about how much the child is learning from the sheets.

In order to satisfy himself he gives supplementary tests. . . . The Teacher's desire to give tests arises out of an actual need of his own to re-establish his confidence as a teacher. This is one way in which the Teacher tries to ensure commitment. The Teacher whose knowledge of his children's progress arises only from his contact with their written work must inevitably feel he has lost that contact when most of what he sees are simply answers. . . . He no longer feels that he knows that they are learning what he is supposed to teach them: mathematics. For this Teacher, to be a teacher at all means that he must feel he knows how much of the work his children have learned. . . .

Now let us turn to the child faced with a set of blanks to fill in on a page, and the knowledge that when he has done that he can go on to the next page, and the next one. . . . The red sheets are often very simple to start with; most children, set off to fill in the blanks, which they do with gusto, until they come to one they are not sure about. Then they may go back and read what they were supposed to do. It is easy to say that they *should* read the instructions, but how many times do we read properly instructions for operating a new machine, washing powder, toy, paint, before we actually start on the job? And how far are they meaningful before we actually start? An alternative course of action is to ask a friend or to ask the teacher what to do. The choice depends on the ethos of the classroom, how silent, how quiet, how noisy, and on one's status with the teacher. There is real need here for the teacher to teach the child how to deal with instructions but instead it is too often found easier simply to tell him what to do. 'You turn this round and measure this bit and this bit. . . .' 'You take these numbers in turn and work out the sum each time with the number instead of x'. . . . Very often a sentence is all that is required for the child to be able to start filling in the blanks again. . . .

I want to leave that for the time being and return to a picture of the child facing the red sheet with blanks. He lives in a society where competition is regarded as a good thing; all his school career he has been encouraged to compete with his friends. . . . What then is the child's natural reaction to a sheet with blank spaces to fill in? He fills them in as quickly as possible, and preferably more quickly than his friend. When he has filled in that sheet he can get another and another. . . . Some children who have already begun to ask questions about the point of mathematics or school will see in this an infinite path to boredom; others will see it as a challenge: who can get the most sheets done in the shortest possible time. Quicker if you miss the cards/games/apparatus out. Quicker still if you copy down the answers from someone else. . . . I want to emphasise the pressure of speed felt by the child doing modular work. This pressure arises quite naturally and spontaneously without any overt encouragement from the teacher; it is the immediate reaction to a set of blank spaces waiting to be filled. When the teacher has also to assess the child partly by the length of time he takes over a module in comparison with everyone else in the class, it is clear that the strong feeling for speed can become a danger in that it obscures the work being done. . . . I am illustrating the classroom situation in which the mathematics modules are being used by a particular Teacher. This is the Teacher whose ideology, whose particular structure for teaching is governed by thoughts . . . which are mostly concerned with the work-to-be-done, and only peripherally or incidentally with a child. A Teacher, who only knows of access to a child through answers given, is driven, because of the blank-space-filling-format, to give extra tests and examinations. He cannot see what he, himself, can do as a *teacher* in this individualised learning context. A teacher who believes that the sheets are self explanatory will not expect the children to ask many questions. . . . Thus he replies either with a minimum of explanation, a reiteration of what is written down, or he tells the child what to do to get the answer or, assuming the child has not read the instructions, simply says 'go and read the instructions'. Such actions by the Teacher indicate to the child the importance of the *answer* in his mathematics lesson; there is little or no *conversation* about the work and the child's feelings about the completion of the sheet are reinforced. . . . The emphasis is on getting through the module.

As he faces his sheet in such a welter of feeling, the child, . . . has little time to read the instructions in a way which we would describe as 'carefully'. He looks at the first half of the sheet and searches for clues on how to fill in the blanks. He looks and finds a visual pattern, he checks his pattern with another example, if they fit, he proceeds and fills in as many spaces as he can. He barely notices the carefully underlined or capitalised phrases which remind him succinctly of what he is doing. He is anxious to get finished and proceeds as fast as he can until he gets stuck. When he gets stuck and asks for help, he becomes willing to receive only the necessary minimum of explanation (as we all are) which will lead him to the answer to be put in the blank space. It is not true to say, on the whole, that he wants to be told the actual answer when he gets stuck. Such a desire is comparatively rare and is consequent on other kinds of pressure in the classroom. He needs information to help him extend his pattern which has worked successfully for all the others, or alternatively information to make him discard the pattern. . . .

The format of the sheets; the explanations, examples, blank spaces, lend themselves to a patterning process which we may or may not call understanding. The ethos of the classroom accentuates or hinders this process but the classroom I am using to illustrate my points will most certainly accentuate the patterning process to the denigration of understanding. . . .

(In) such a system . . . (it) will be inevitable that the Teacher finds all his time consumed by administrative affairs . . . because he does not know what else could possibly fill his time. . . .

Such a teacher is in definite and destructive conflict with his previous ideologies. He despises the system because it so clearly dispenses with his especial talents; and yet he worries because in some obscure way the children seem to be easier to manage, and he has no idea why.

Exercises

1 Morgan observes that for the child working with modules the pressure of speed 'is the immediate reaction to a set of blank spaces waiting to be filled'. Her observation is similar to Smith & Pohland's finding that the CAI programme seemed to encourage competition among pupils to see how fast they could work. Is the pressure of speed an inevitable accompaniment of attempts to individualise instruction? Discuss how an evaluator might investigate the effects, both beneficial and harmful, of the pressure.

2 Find out more about the Scottish Education Department individualised learning project. Which of the approaches to curriculum development outlined in chapter 5 was used in the project?

3 Martyn Cundy (1976) evaluated the Caribbean Mathematics Project by composing an historical account of the project, studying the project's materials and preparing a critique of their mathematical and pedagogical soundness, visiting schools to observe classes and to talk to teachers and principals, giving to teachers questionnaires about their attitudes toward the project and their objectives for mathematics teaching, and to pupils a test of their knowledge and understanding of mathematics and an inventory of their attitudes toward mathematics, and visiting teachers' colleges to study how the materials were used in teacher education. Cundy presented an island-by-island survey of the project's status and then evaluated its achievements in terms of its overall strategy of development, its use of international cooperation, its cost effectiveness, its observable effects on pupils and teachers, and its potentialities for the future. Cundy's study was much more comprehensive than that of Morgan, which dealt with only one facet of a project, namely, the affective consequences of its approach to individualisation. Both studies, however, can be classified as illuminative evaluations of projects that used a formative approach. Which study seems more likely to have provided better information on (*a*) how the project was influenced by the various school situations in which it was used; (*b*) what it is like to be a participant in a project class? Why?

4 Crawford attempted to ask the pupils unbiased questions about the Fife Project, but the very fact that he was soliciting opinions may have led the pupils to infer that he wanted to hear favourable comments about it. What methods can evaluators use to obtain honest reactions from pupils?

5 An evaluation may aim to produce information which can be generalised or information which is 'project specific'. Its methods may be widely applicable or of restricted validity. Again, it may seek to set

out for administrators or teachers either 'generalised conclusions' or 'generalisable descriptions'.

Compare the various studies we have summarised. What did they seek to produce?

7. The evaluation of projects based on the integrated-teaching approach

To the extent that a curriculum project based on the integrated-teaching approach attempts to do away with the boundaries between mathematics and other disciplines, it will encounter problems in implementing its programme with teachers whose own education has been conducted largely within these boundaries. And to the extent that such a project attempts to develop mathematics out of real situations, it will encounter problems with teachers whose mathematical education has not prepared them to see opportunities to devise a mathematical model of a situation. Projects based on the integrated-teaching approach, therefore, have faced greater problems of dissemination and teacher education (and re-education) than those based on other approaches. These problems should also have been considered in devising evaluations of the projects. As the following example shows, however, the agency that funds the project may be interested more in whether the project 'works' than in whether it can be disseminated.

7.1. *The USMES evaluation programme*

The principal evaluation studies of Unified Science and Mathematics for Elementary Schools were undertaken by Mary Shann of Boston University and her associates under grants from the National Science Foundation, the project's sponsor. USMES had conducted its own informal formative evaluations, but NSF sought an independent, more formal, and more summative evaluation. Shann and her associates attempted an evaluation of USMES during the 1973–4 school year, but failure to receive funding until mid-year resulted in serious losses of data on pupils' performance. The 1973–4 evaluation concentrated on visits to USMES schools, interviews with teachers and other school personnel, and questionnaires mailed to teachers. It became, of necessity, something of a formative evaluation and a pilot study for the following year's evaluation.

NSF's concern in funding the independent evaluation studies was to obtain evidence regarding 'proof of concept', that is, whether or not the programme was actually doing what is claimed: increasing pupils'

abilities to solve real problems. Shann had proposed a broader evaluation study in which questions of how the programme was being used,
how teachers were trained to use it, and how it was disseminated would
also have been addressed. Support for activities to investigate these
questions was not provided, however, and the 1974–5 evaluation study
(Shann *et al.*, 1975) was restricted to questions of the programme's
effects on students with respect to their problem-solving abilities, their
achievement in school subjects, and their attitudes toward various
aspects of the curriculum.

Forty classes using USMES were selected so as to represent a variety
of grade and socio-economic levels, and curriculum units ('challenges')
being studied. Each USMES class was matched as well as possible with a
class from a school not using USMES but from the same geographical
area.

Most of the USMES classes worked on only one unit during the year.
The time spent on USMES differed markedly from class to class, but the
average time spent on a unit was one and a half hours a day, three days a
week, for twelve weeks. Most of the time for USMES was taken at the
expense of science, and to some extent language, arts, and physical
education. The time for mathematics instruction did not seem to have
been much affected. Observations of classroom activity showed that,
although teachers dominated the activity about one-fifth of the time in
both USMES and comparison classes, the pupils in the USMES classes
spent more time in behaviour judged active, self-directed, and creative.

Interviews with teachers, pupils, and administrators showed that
pupils enjoyed USMES and looked forward to using it. Teachers and
administrators expressed considerable support for the USMES philosophy of 'real problem solving' and for an integrated approach to
teaching. Interdisciplinary teaching was made difficult in practice,
however, by departmentalised programmes, rigid time schedules, and
the teachers' limited backgrounds in different disciplines, especially in
science. Furthermore, some of the USMES challenges were perceived
by the children (and sometimes by the teacher) not as posing a problem
but as providing them with a series of unrelated activities. This
perception was shared by some administrators and teachers who expressed concern with whether USMES was teaching problem solving or
'gimmicks'.

USMES teachers and pupils had no difficulty in identifying the
mathematics concepts learned in USMES, but they could not identify
much science in the scheme. The teachers tended to select the chal-

lenges they felt most comfortable teaching, and consequently most of the units they used dealt with social science questions to which mathematics could be applied. The teachers tended to rely on the USMES workshops or word-of-mouth, rather than the teachers' manual, in learning to use the programme. Other USMES material, such as the 'how-to' cards and the technical papers, were not used much by either teachers or pupils, and Design Lab usage declined in 1974–5 from the previous year.

A careful study showed no statistically significant differences in mean achievement between the USMES and the comparison classes on tests of reading comprehension, mathematics, science, and social science. The evaluators interpreted this finding as indicating that USMES usage was not adversely affecting basic skill development. USMES and comparison classes did not differ in their attitudes towards mathematics, towards science, or towards activities embodied in the USMES approach.

Investigation of the question of whether USMES was improving problem-solving ability was hampered by a dearth of measuring instruments. Shann and her associates devised two measures that attempted to simulate a real-life problem situation. They were termed the 'Playground Problem' and the 'Picnic Problem'. Shann *et al.* (1975, pp. 129–30) describe the two problems as follows:

> The Playground Problem . . . required that students develop a plan for a playground which would serve children in their school and/or neighborhood. A catalog of equipment, cost data, and measuring instruments were given to the students along with the information that they could spend up to $2000.
>
> The pre-test, post-test control group design used in the evaluation necessitated that a parallel form for the Playground Problem be developed, since retest results from such a unique test would be affected by memory factors. To answer this need, the Picnic Problem was developed. This test challenged students to develop plans for a class picnic. The students were provided with a photograph of various foods available to them and a map drawn to scale which included the locations of their school and three park areas as possible sites for the picnic. Along with measuring instruments, the students were given cost data and the information that they could spend up to $25. They were to assume that 25 students would be going on the picnic, and that a school bus would be provided for their transportation, free of charge.
>
> Neither the playground problem nor the picnic problem satisfied the developers' concern that these tests meet all the criteria for 'realness'.

The tests were simulated problems whose solutions would not have immediate, practical effects on students' lives. Nevertheless, . . . the vast majority of students tested with the Playground and Picnic tasks were motivated to accept the problems. In that sense, we can say the tasks were meaningful to the students.

Another of the developers' criteria for 'realness' is that real challenges are 'big' enough to require many phases of class activity for any effective solution. The Playground and Picnic Problems did not meet this criterion. In the interest of observing reasonably larger samples of children we had to abbreviate test times to approximately one hour.

Despite these limitations, the Playground and Picnic Problems have other important features in common with USMES-styled, real problems: they have no 'right' solutions; they have no clear boundaries; they require students to use their own ideas for solving the problems; and they elicit group efforts toward the solutions to the problems.

Shann developed a scoring guide for each problem to help standardise administrative procedures. Excerpts from the scoring guide for the Playground Problem are represented below (Shann *et al.*, 1975, pp. 101–6).

Organization

1. *Selection of Children*
A random sample of five children should be picked from each control class and each USMES class in your school. . . .

It would be best for you to pick the children yourself, but the teacher can also make the selections if correct procedures are used. The easiest appropriate method is to write the names of each child on a piece of paper, throw each piece in a hat, and then select five.

2. *When to Administer the Playground Problem Test*
This can be a critical factor. . . .

Try to run your test sessions at approximately the same time of day – that includes the control classes as well as the USMES classes. the recommended time of day is as close to the beginning of the day as possible. . . .

3. *Where to Administer the Playground Problem Test*
In preparation for the test, you should locate a suitable open area near the school. . . . This area should be the same for all groups of children in the same schools on your sample list, both USMES groups and control groups.

4. *Materials to Accompany Test Administration*

Prior to the testing session, you will need to gather together the following items:

Observation Equipment	*Tools (in a cardboard box)*
Observation form	50 foot tape measure
Tape recorder and	Yard stick
blank cassette	Ball of string
Watch	Large piece of paper
	Tri-wall (to use as hard surface for
	drawing plan)
	Felt tip pens
	Pencils
	12″ rulers
	Catalog of playground equipment
	Scrap paper
	Scissors

Instructions to the children

Soon after arriving at the open area, you should give the children the following instructions and you should record them on tape:

> 'Let's suppose this area was going to be made into a new playground for the children in your school.' (Indicate clearly the limits of the area.) 'How would you plan this playground?'
>
> 'Here is a catalog of playground equipment which could be bought. If you had $2,000 to spend, which equipment would you choose?'
>
> 'Please work together to decide which equipment should be bought. Draw a plan of the playground on this piece of paper showing where the equipment would be placed.'
>
> 'You have forty minutes to work together to make your plan. Here are some things you may use if you want to.' (Hand one child the box containing the tape measure, pencils, etc.) 'Remember, you can spend up to $2,000 on equipment.'

DO NOT GIVE THE CHILDREN ANY SUGGESTIONS AS TO WHAT OTHER CONSIDERATIONS THEY SHOULD KEEP IN MIND. . . . The instructions should be as similar as possible for the USMES groups and for the control groups. Any evidence of intentional or unintentional bias unfortunately results in invalidation of the test session.

Let the children know that they will have forty minutes to figure out their plan and draw it on paper. Tell them that at the end of this period, you will ask them questions about their plan, and that their answers will be recorded on tape. . . .

Question period

This period during which the children explain their plan and outline their reasoning should be tape recorded in its entirety. The children's presentation may be *up to ten minutes long*. . . .

It is very important to remember that the questions you ask the children and the procedures you use in soliciting their answers MUST be as similar as possible for the USMES groups and for the control groups. Again, any evidence of bias may invalidate the results. . . .

First question series (Directed to the entire group*)
- 'How did you do?'
- 'Was it fun?'

Second question series (Directed to the entire group*)
- 'Explain your playground plan.'
- 'Why did you decide to buy (4) pieces of equipment?'
- 'Do you know how much the equipment you have chosen will cost?'
- 'Why did you decide to put the swings over here? The slide over here?'
- 'What kinds of information did you need to help you make your decisions?'

Third question series (Directed first to the entire group, and then to each child in turn who has not yet responded)
- 'Were there any other important factors you had to consider in making your decisions?'
- 'Is there anything anyone would like to say before we finish?'

While it may be necessary to structure the children's report by asking questions, you as the observer should not suggest rationale to the children by means of your questioning. For example, if there has been no mention of safety factors or indications that the issue of safety has been taken into consideration, the observer should not bring it up during the tape recording.

The playground problem does not have one solution. However, in the playground problem, a certain approach to problem solving is valued. An excellent response to the playground problem would include:

1. Measurement or calculation of available space.
2. Meaningful use of measuring equipment.
3. Careful consideration of types of playground equipment chosen.

---

* When the question is directed to the entire group make sure that everyone talks who wants to, not only the 'spokesman' for the group. Be sure they talk one at a time so that it is easy to understand what is being said.

4. Comparisons between size of equipment as listed in catalog and space available on playground area.
5. Consideration of budget limitations.
6. Accuracy in drawing lay-out of proposed playground.
7. Consideration of human elements such as safety and aesthetic appeal.
8. Logical and clear presentation of rationale.

Unfortunately, despite the effort expended to construct reasonable simulations of real problem solving, the USMES pupils did not differ from the comparison pupils in any aspect of their work on the problems. Of course, one can always complain that the tests were not 'real' enough. Evidence remains to be found, however, to support the USMES developers' and teachers' contention that the programme does improve problem-solving skills.

A final issue regarding USMES is raised in the following excerpt from the evaluation report (Shann *et al.*, 1975, pp. 251–3).

> [This report] . . . documents strong support for the USMES project, both in its conceptual design and in its classroom implementation. The sources of these evaluations have included teachers and principals engaged in the use of the USMES program.
>
> In apparent contradiction, several of these same subjects engaged us in confidential interviews at almost every site we visited and communicated to us attitudes of disappointment and disenchantment with the USMES program. We estimate that these confidential interviews and their negative content represent about 50% of the teachers and administrators using USMES on the local level. Their comments are representatively paraphrased as follows:
>
> > At this school, only one-third of the original USMES teachers are still using USMES. Some of our 'drop-outs' are quite bitter.
> >
> > (a principal): While I am very supportive of the USMES program, I can't get any of my teachers to use it. . . .
> >
> > 'X' is an exemplary USMES teacher, on paper, but in fact he never uses USMES in his classrooms and his logs are pure fiction. . . .
> >
> > Don't say anything bad about me back in Boston, Tell EDC what a good job I am doing so I can go back to Boston again next summer.
>
> If these complaints are interpreted as critical of the program itself, they clearly conflict with the more objective and carefully analyzed information reported throughout this study. Nor did we, as participating interviewers, understand their thrust to be in this direction. What we did hear were indications of a serious morale

problem among the USMES trained teachers and principals on the local level. And this on two counts: (1) Principals supporting USMES in their schools feel they are not in communication with the program's officers, especially in regard to the changes in direction which have occurred over the past three years. (2) The pivotal teacher representing USMES in each local school frequently is a personality who *says* all the right things about USMES but does not *do* USMES in the classroom. His leadership is seen as more persuasive than honest. In these cases, he is not viewed either as a teacher of integrity or as a genuine implementer of the USMES philosophy. This representative figure, however, 'personifies' USMES to his colleagues on the local level; to all those within his sphere of influence, his failings become representative of the value-estimate of the entire USMES program.

Because of our assurances of confidentiality, and because we designed no instruments which would test these injections against objective evidence, the observations presented here must remain on the level of hearsay. It must remain the task of some future study to determine if, in fact, such a morale problem does exist; to examine the extent of its spread; to estimate the depth of its effect; to gauge the degree to which it endangers the future of the USMES development; to identify its causes; and to suggest possible methods of correction.

Our research indicates only that the USMES program, in comparison to other NSF curricular programs, has a distinct philosophy which relies heavily on the personal quality of its teachers. 'USMES is more a philosophy than a set of materials.' This factor makes the USMES method of selecting, motivating and training personnel more critical and more vulnerable than is true of other NSF programs.

The USMES project was difficult to evaluate precisely because it aimed so high. That is, it claimed to deal with skills and abilities that everyone thinks are important but that very few people would claim to know how to measure, and it made demands on teachers that few people are equipped to meet without support. The support system that USMES provided through its workshops for teachers and its communications with them appeared to be critical to the success of the programme: when it was withdrawn as funding for USMES was reduced in 1976 and 1977, the use of USMES material declined rapidly.

Exercises

1 The 'morale' problem identified by Shann *et al.* in the preceding excerpt appears to have arisen in part because the project depended

heavily on team leaders, and at least some of these leaders took advantage of various perquisites and did not have the respect of their team.

(a) Do projects following the centre-periphery model for dissemination and using team leaders to keep the project functioning at the periphery necessarily face such a morale problem?

(b) Design an evaluation study to ascertain the extent and effects of the USMES morale problem.

2 Attempt to devise an alternative to the picnic and playground problems which could be used for testing purposes.

3 How fair a simulation of the USMES project activities are the testing procedures devised by Shann? How likely are they to assess what students have learned whilst participating in the USMES programme?

7.2.   *Other studies of integrated-teaching projects*

The Mathematics for the Majority Project was one of the first UK projects to incorporate provision for formative evaluation from the outset. Since the project sought to aid children of 'below average' mathematical ability, the evaluator was first asked to discover more about their teachers and the kind of help they needed. Kaner (1973) found the position differed throughout the country, but that over half the target students in grade 8, and almost as many in grade 9, were taught mathematics by non-specialist teachers: data which indicated the task confronting the project team.

Most of the evaluative effort of MMP went into pilot-testing trial versions of the project's teacher's guides. An elaborate scheme for drafting, testing and rewriting the guides was devised; one timetabled to take some two years. Since team members typically served for two years and the project itself lasted only three, the implication is obvious: few guides could be produced that made use of the firsthand experience of field trials. The large number of prepilot schools also retarded the production of the guides. Moreover, the pilot schools found it difficult to supply the monthly reports and information requested by the evaluator. Case studies of some schools and visits to others did, however, result in valuable feedback to the central team.

One request was for pupils' materials to accompany the teacher's guides, a need to be met by the Mathematics for the Majority Continuation Project. The MMCP had a team of three evaluators who used questionnaires and visits to schools as the bases for formative evaluation. Unfortunately, the project never came to grips with the

problems of dissemination, and no summative evaluation was undertaken.

MINNEMAST, which began in 1962 and continued to 1970 (at a cost of
five million dollars), was one of the first projects to attempt to
coordinate science and mathematics teaching in the elementary school.
It was the subject of an evaluation by Hively *et al.* (1973) who devised
the notion of 'domain-referenced' achievement testing. This implies the
construction of rules for generating sets of equivalent test items
representing clusters of related concepts and skills. An evaluator was
assigned to each writing team as a consultant in educational objectives
and to help them specify domains within which test items could be
written. What in theory sounded plausible proved to have limitations in
practice. Indeed, when reading Hively one can almost hear the evaluators' cries of pain as they disappear beneath a mountain of data and
their careful advice goes unheeded.

Exercise

Projects based on the integrated-teaching approach seem to have
encountered acute problems of dissemination. What evaluation
methods are especially suited to the study of disseminatory activities?
Design an evaluation study of the dissemination of a suitable project.

8.   Concluding observations

The evaluation studies surveyed in this chapter have been as
varied as the projects they have evaluated. Some have been conducted
by insiders, some by outsiders. Some have been focused on one aspect
of a project, others have been comprehensive. Some have generated
numbers, others ideas. Some have determined the fate of a project,
others have been ignored. Some have been thorough and systematic,
others cursory and haphazard. In general, the studies have been
empirical in nature. Yet, of course, there is also a demand for
evaluation with a theoretical orientation: a retrospective, theoretical
appraisal of a project's premises, objectives and methods.

The art and science of educational evaluation is still in a primitive
state, and it is not surprising that its practice does not follow well-worn
paths. Nevertheless, evaluation of sorts is constantly taking place, and
decisions are being made on the basis of so-called evidence. The need to
develop improved methods and to inculcate new attitudes is, therefore,
obvious. For, if we do not do so, we run the risk that those programmes

and initiatives which are best fitted to evaluative procedures as they currently exist will thrive at the expense of others. The challenge, then, to evaluators of mathematics curricula of the future is to build on the best of present practice without forgetting the lessons to be learned from the past.

8

Lessons for Today and Tomorrow

It is now almost thirty years since the 'modern math' reforms began. Today, much of the enthusiasm of those early days seems to have been lost. There is a feeling that 'change' has been overdone: 'innovation' is going out of fashion.

That 'innovation' should go out of fashion is not necessarily to be deplored: few would wish to see the like of the 1960s again. Then, too much was attempted in too short a time with insufficient thought concerning aims and possible effects. Projects were looked on as 'experiments' and many included that word in their title. Yet an educational system is not like a chemical laboratory where a failed experiment means merely that certain chemicals and apparatus have been expended. A failed experiment in education can mean deprived pupils and dispirited teachers, and parents who will look askance at further proposals for change. More than money can be lost when a project fails.

Much current discontent undoubtedly springs from the fact that the *practical results* of such an enormous expenditure of labour and commitment have been relatively insignificant. The problems remain – many Johnnys still cannot add! Indeed, our ability to help students attain mastery remains low. Presumably, today we understand much better the complexity of the problems to be overcome, but is that the only outcome of twenty years' work?

To pose such a question is to ask for an evaluation of curriculum development in a different sense from that we described in the previous chapter. No longer are we evaluating a single innovation; but now we enquire about the outcomes of curriculum development as the historical process witnessed after World War II. Yet such considerations would seem particularly appropriate to the last chapter of a book in which we have tried to acquaint the reader with various aspects of curriculum development. In this chapter, then, we have two aims. First, we look critically at the reform period in retrospect, its achievements and failings, and the lessons to be learned. For this purpose we do not have recourse to many analytical studies; in particular, no comprehensive, international analysis of the reforms has been written to date, nor could such an undertaking be readily accomplished in the near future. In this area the frontiers of research are soon reached!

We then consider what we believe should be the teacher's and student teacher's commitment to curriculum development: for this book was written in the conviction that teaching and curriculum development are inextricably linked. We hope that this will indicate how use might be made of what is to be found in this book, and that these closing remarks will help counterbalance the many criticisms which we make and so leave our readers with a more positive outlook.

1.   On the evaluation of curriculum development
1.1.   *The development of curriculum development*

Since the 1950s curriculum development has itself developed; moving from small beginnings to the prosperity of an academic, even scientific, reputation. In so doing, ideas, orientations and approaches have been changed. What prompted these changes? How are they to be understood and explained? Did they happen more or less accidentally, reflecting new findings in related disciplines? Are they due to the gradual moulding and forging of new ideas in the 'furnaces' of classrooms, educational systems and societies? May we understand them as steps forward in a continuous progress, each being based on what has gone before?

In previous chapters we have provided examples which would support an affirmative answer to each of these questions. No doubt the emergence of Piaget's theory inspired a new 'round' of projects; the behaviourist projects in the United States owed much of their success to their ability to match the demands of a commercially oriented market; and in the same country the Cambridge Conferences showed that curriculum development can be pursued as a process of deliberate progress.

Of course, US experiences differ from those elsewhere. In other countries smaller financial resources and the heavier hand of tradition usually restricted the range of approaches to be found. Yet most of the approaches and problems encountered elsewhere can be identified within the US and for that reason a study of the evolution of curriculum development there would seem of international interest.

As we have seen in chapters 5 and 6, the basic 'approaches' of curriculum development emerged in the US one after the other. The concerns of the individual were emphasised by Dewey and his school; those of society by Bobbitt and his followers; finally, those of the discipline by the proponents of the New Math. Theoretical frameworks were supplied to support each of these aspects, although in the case of individual-centred developments a scientific foundation based on theories of cognition post-dated the initial postulation by many years.

Once it was recognised that each of these concerns was valid, attempts were made to find a comprehensive theoretical structure. Establishing this proved an extremely difficult task, the only theory to achieve this integration (on a rather high level) being that of Bruner. Yet this was a sophisticated approach which proved difficult to translate into practice. The demand for easy-to-handle classroom tasks often resulted in embodiments being divorced from their structural aspects and reduced to more traditional, whimsical applications of mathematics. Moreover, the need to preserve a consistency of embodiments with a particular formalised language of their own, frequently resulted in a tendency to hide structures rather than to reveal them.

Nevertheless, the structuralists set new standards concerning the balance and integration of the constituent aspects of curriculum development on a scientific level. (In doing this, however, they perhaps posed more problems than they solved.) One might have expected such standards to have been at least maintained by later developers but, as we know, they were not. The reason lies in the highly contradictory potential of aims, depending on the different constituent aspects of curriculum development. The equilibrium of demands realised in the structuralist approach was lost when, in the middle of the 1960s, social obligations and engagement claimed more consideration within the curriculum than the structuralist approach had conceded. The conflict of individual orientation *versus* social needs was to be resolved in an elegant way: 'to learn learning' satisfied both demands, but it was a somewhat unsubstantial compromise.

The existence of contradictory claims to be mediated within the curriculum led of necessity to problems of weighting. And this, of course, meant that curricular decisions became related to underlying philosophical and political convictions. Politicians have always understood the political dimensions of curricula, but often their interests have been allowed to pass without comment. Events in the late 1960s, however, focused attention on the political aspects of curricula, and it was maintained that curriculum development should contain within it the ability to spot concealed intentions. In particular, it was argued that the objectivity and disinterestedness claimed for pure science and mathematics, had often served as a façade behind which other interests were pursued – what Philip Jackson referred to as the 'hidden curriculum'.

To some extent the formative and, even more, the integrated teaching approach were inspired by this social/political dynamic. This meant, however, that less emphasis was accorded to the claims of the discipline concerning extent, selection and sequencing. The curriculum became determined more by method, and less by mathematical content (which risked becoming the subject of arbitrary decision-making). The integrated-teaching approach strengthened this tendency. But if mathematics is to be reduced to an instrument for problem solving, then a vast and ill-defined area is opened up and the question of the legitimation of content takes on a new emphasis. Yet the major projects associated with this approach still seemed to rely on somewhat accidental procedures to determine content areas. The question of balance remained unresolved.

Exercises

1 In what sense is it correct to describe 'innovation' in the sixties as a 'fashion'? What place is there in curriculum development for 'fashions' and is their effect necessarily to be deplored?
2 Give examples of how national traditions have 'restricted the range of approaches to be found' in curriculum development.
3 There was 'a tendency to hide structures rather than to reveal them'. Investigate this assertion.

1.2. *Curriculum development and the reality of school practice*

In the previous section we briefly hinted at some of the theoretical problems which arose from the basic premises held by developers. Yet, in surveying the events of the past twenty years, it is the practical problems concerned with implementation in the classroom which most obviously catch the attention. No doubt, most of the disenchantment with modernised curricula arose from practical difficulties, possibly even organisational trifles, than with deeper, theoretical and philosophical, considerations. It is often asked to what extent innovators are aware of the day-to-day difficulties posed by life in a real school. Certainly, one of the lessons to be learned from the last twenty years is that when innovation clashes with classroom reality, it is the former which is usually forced to adapt. We recall how early projects applied the 'technological' or 'R–D–D' model of innovation (see chapter 5, section 4) which entrusted curriculum development to specialists who attempted to create curricula as far detached as possible from the imponderables of the classroom. This innovation model, bound to the 'New-Math' and 'behaviourist' approaches, lost its justification when, on the one hand, it was acknowledged that pedagogical problems could not be ignored, and, on the other, that behavioural learning could not encompass the whole curriculum. The conclusion was that no totally predetermined curriculum was possible. (This does not mean that behaviourist methods cannot be successful on a limited front. It is only when their neatness and effectiveness tempt us to seek wider educational goals that troubles arise.)

With the emergence of the structuralist and formative approaches to curriculum development, the learning process itself, and with it the teacher, regained their central place in curricular thought. Indeed, the outstanding lesson of twenty years of concentrated – indeed, frantic – curriculum development is the crucial role which the teacher has to play. No matter how outstanding a project's teams or materials may be, the success of its work will ultimately hinge upon the receptiveness and adaptability of the classroom teacher. Or, to put it another way, success depends highly on innovation models which ensure the appropriateness of materials by means of early teacher participation in the developmental work itself. Where, as in Britain, curriculum development lies within the teachers' domain, interaction between the levels of development and application will automatically be greater than is to be found in other innovatory procedures.

However, such collaboration is still no guarantee that developments conceived with the teaching force in mind, or even within that force, will in the end actually match it. The teachers involved in the preparation of teaching materials will soon tend to occupy an exceptional position, and will cease to be representative of the force as a whole. This tendency, to which little account would seem to have been paid in the innovation strategies of even the more recent projects, produces an ever increasing gap between 'reality' within an experimental school and within a more typical one.

Further initiatives are likely to cause the gap to widen even more: in the US during the 1960s and early 1970s experiences gained from projects were precipitately translated into new approaches. Yet there was hardly a school that could amend its aims and practices as quickly as could the developers. The gulf between projects and reality widened. Thus, for example, the introduction of a Madison Project curriculum in a school that had followed a behaviourist programme seemed doomed to failure. In spite of counselling, the difficulties met by teachers and pupils were immeasurable. It is not surprising therefore that the Madison Project although greatly esteemed by curriculum developers made little actual impact on schools.

Exercises

1 'When innovation clashes with classroom reality, it is the former which is usually forced to adapt'. Discuss and exemplify.
2 How can one attempt to ensure that teacher-innovators do not, in the eyes of less enthusiastic colleagues, 'disappear over the horizon'?
3 Find examples where there have been mismatches between the innovators' view of the teacher's role and that held by the teacher.

1.3. *Curriculum development and social reality*

Developmental work can, as we have seen, be on a small scale and intended for restricted, local use. Yet curriculum development as we have described it is more often on a large scale. It is not cheap and it demands a rich amalgam of skills and expertise. Usually, then, it is conceived 'for others' and wide dissemination is one of its major objectives; indeed there are often commercial (and prestige) pressures to make it the principal one.

Dissemination, however, raises practical problems in a dimension additional to that of the classroom: that of the social and cultural field into which the innovation is to be implanted. Differences between, say, industrial and rural backgrounds, or poor and rich districts can create problems within a country, yet they grow significantly if the dissemination crosses frontiers. Mismatches between the social and cultural backgrounds of the donor and recipient can prove disastrous. The point is doubly important when one realises that only a few countries can maintain independent curriculum development. Many countries, in particular the majority of developing ones, rely heavily on the large-scale importation of modern curricula. In other countries, for example, West Germany, there have been very few indigenous projects. Curriculum development has been based on the eclectic adoption of foreign research and developmental work.

The absence of significant major projects in West Germany (with the exception of Bauersfeld's Alef Project) results largely from the peculiarities of the German school system. It is Federal but highly centralised within the various Länder. Changes in the curriculum are implemented by administrative fiat and laid down in an obligatory syllabus. The syllabuses (different for each land) are given in some detail, and it is left to textbook writers to translate these into concrete teaching materials. Accordingly, textbook writing occupies a corresponding role to project work elsewhere. There is a significant difference, however: the textbooks are tied to the syllabus. Innovation is reserved for administrators.

These conditions, together with the conservative educational policy followed in the post-war years, had effectively frozen innovation until the wide-reaching reforms decreed in 1968 produced a sudden thaw. Teaching materials were then hurriedly produced using ideas gathered from a variety of sources including SMSG, UICSM and Dienes. The outcome was as unfortunate as the circumstances should lead one to expect. The so-called 'bureaucratic' innovative strategy, like the technological model, ignored the teacher's possible contribution. The new syllabus was introduced without any preparatory in-service training and without any serious attempt to involve teachers in the reform process. 'Set theory' became the trade-mark of the new syllabus: it was understood neither by pupils, by parents, nor even by many of the teachers (who had to learn it from their pupils' texts). Failure and resulting complications followed. The reform, inspired by foreign examples, had ignored the reality of conditions at home.

The developing countries have traditionally had to look overseas for their teaching materials, their educational systems usually being too small to be self-sufficient. The way in which this transfer of teaching materials takes place varies considerably and tends to evolve along with the educational system of the recipient country. To make the point clearer, let us look briefly at the history of the transfer of British materials to that one-time 'developing country', the USA.

America's most popular eighteenth-century arithmetic text was Thomas Dilworth's *The Schoolmaster's Assistant*. This was simply a reprint of the English text: but, to give it some local appeal, the name of Nathaniel Wurteen, 'Schoolmaster at Philadelphia', was added to the list of fifty people who endorsed the work in its Preface (see NCTM,

1970). Dilworth's book was later ousted in public affection by Daboll's, again derived from an English text, but now including in its title those magic words 'adapted to the United States'. The extent of the adaptation appears to have been the inclusion of some questions on dollars and cents; most questions, however, still used pounds, shillings and pence! By 1850 the English domination of the textbook market had come to an end. The emergence of indigenous authors, and the widening differences between the two educational systems virtually ruled out the transfer of materials on a large scale.

Transfer of materials has not, however, completely ended, for recent reforms have led to increased interest in the work of educators elsewhere. Thus books produced by some British projects have had an appreciable sale in the USA; in particular, a number of schools have used an adaptation of SMP materials. However, transfer has taken place mostly at the level of ideas and it has been two-way; classroom transfer has almost ceased to take place. However, at the university level, widespread transfer still persists. Here the mathematics is more culture-free, there are fewer nationalistic tendencies, and the market (particularly for postgraduate texts) is smaller.

Looking at these and other examples of transfer we see it is possible to distinguish various levels: there is the straightforward acceptance of materials produced elsewhere; there are small-scale adaptations (often carried out in the exporting country) and large-scale ones, invariably carried out in the recipient country; and there is the assimilation of ideas produced in other countries. As a country's educational system develops so it would appear to progress through these various levels.

The phenomenon of large-scale adaptation is a recent one. It arose from the wish of developing countries to keep abreast of developments they saw elsewhere, coupled with the fact that they lacked the finance and expertise to devise materials *ab initio*.

Paradoxically, the newer materials (other than those firmly in the 'New-Math' camp) have usually proved more difficult to transfer than traditional texts. Hall & Knight's *Elementary Algebra* (1885), was a textbook on which 'the sun never set'; for it was used throughout the British Empire. It was, however, so culture-free that it could be used to equal effect anywhere. On the other hand, the SMP, say, has laid great stress upon the need to draw out mathematics from children's day-to-day activities: thus, for example, one of its series bases its introduction to coordinate geometry on the game of 'Battleships'. This, one suspects, is not an approach which would mean much to a girl in land-locked

Lesotho or Malawi. The answer would hardly seem to lie in teaching children to play 'Battleships' in order to introduce coordinates! It is necessary to look elsewhere for suitable background experiences and if none exists to consider seriously whether one should not postpone the introduction of coordinate geometry and turn to some other aspect of geometry arising more naturally from the child's environment. This, however, would cause vast problems of adaptation within a tightly structured series.

Even more worrying, however, is the fact that the SMP series was designed to meet the needs and objectives of a particular type of schoolchild in England. True, by a trick of history and because of the inertia of educational systems, children in Lesotho and England can share common examination needs, but can anyone pretend that their educational needs are similar?

Yet, as Cecil Beeby wrote (Howson, 1970), 'No matter how fully the principle [of adapting the curriculum to the society served by the schools] is accepted, those who are responsible for education in developing countries know that, through lack of books, lack of equipment, and lack of adequately trained local teachers, they will, over the next few years, be driven to importing educational ideas and practices that are really irrelevant to their needs.' The results, as the reader of the various reports contributed to the two survey numbers of *Educational Studies in Mathematics* (1978) will observe, have been discouraging and in some instances have led to a return to equally irrelevant traditional curricula.

The need to assist developing countries to produce new materials was recognised internationally. It is indeed one of the sadder aspects of many of the recent attempts to transfer materials and ideas that they were made with the best of intentions. They represented a genuine attempt to provide aid. Yet to be of assistance in such a delicate field as education requires a background knowledge of the societies and systems to be helped that few possess. There is an almost unsurmountable, even subconscious, temptation to impose on others one's own aspirations and patterns of thought. Even when attempts were made to produce original materials specifically for the countries concerned, for example, the Entebbe Project and the UNESCO Arab States Project, the writing teams were dominated mathematically and professionally, if not numerically, by Western authors who lacked any prior understanding of the educational systems of the countries concerned and, more importantly, of the social ethos that was manifested in the schools.

It is easier, however, to criticise past efforts at providing aid than to lay down guidelines for the future. Certainly, it is clear that curriculum development should be led and controlled by educators from the country in which the development is taking place. Only they can have the requisite background understanding of the problems and be able to communicate with teachers in the desired manner. Moreover, only in this way can one maintain continuity of effort: curriculum development does not end with the publication of materials; the hard work is only then beginning. The provision of training programmes and exchange visits with the intention of nurturing and developing a cadre of leaders would seem the first priority. Until there exists a core of well-informed potential innovators within an educational system, any attempts to promote developments within it are likely to fail: there will be insufficient expertise present to 'go critical'. Expatriates may still have a part to play in such developments; they have advice and experience to bring, but if their roles become central rather than subsidiary doubts must be raised about the ability of the reform to sustain itself once they have left the country.

The 'low-level' transfer of materials at the school level is not then to be encouraged. However, it is one of the main tenets of this book, as evidenced by the nationalities and affiliations of the authors, that curriculum development has much to gain through the international pooling of experience and expertise.

(In this connection we note that although international jamborees may serve to introduce people to each other, as proved the case with the authors of this book, they will not suffice, by themselves, to bring the required depth of understanding.)

National differences between educational systems do affect curriculum development greatly, but one must not allow these to restrict internationally based attempts to unravel and solve the problems which developers everywhere meet. The differences of emphasis, strategies and methods to be found within individual countries provide us with evidence that curriculum development cannot be simplistically subdivided into 'national' styles: there will always be much common ground.

Exercises

1 Find examples (possibly from the USMES and MMCP extracts quoted in chapter 6) of material which would be suitable for children from a

particular socio/economic/geographical background, but ill-fitted for pupils elsewhere. (Think not only of 'relevance', but also of motivation, language, etc.)

2   Investigate the history of the transfer of teaching materials *to* your particular country.

3   It has been argued that the three stages of transfer, acceptance, adaptation and assimilation are also to be observed in the way in which individual teachers receive teaching materials into their classroom. Comment upon this and also on ways in which progression through the stages can be expedited.

2.   Curriculum research desiderata

Curriculum development has to date provided a perfect example of the old adage that the more one knows, the more one knows how little one knows. Accordingly, when describing the outcomes of the reform period we are constantly called upon to refer to problems which hitherto have largely remained unobserved. Of course, some of these problems really are new, and result from changes in the provision and orientation of education: they have appeared too recently to have been solved. Yet it is now clear that much research and development work, and considerable ingenuity, will be needed before solutions to even a selection of our problems are to hand. As examples we examine three particular problem areas, before considering more general research matters.

2.1.   *Problems of linearity*

In chapter 6 we saw how projects were influenced by the particular beliefs of their originators concerning the contribution mathematics can make to education in general; what constitutes 'mathematics' and, in particular, what, if any, is its underlying 'structure'; and the weight to be given to research findings on cognitive development.

Yet certain important aspects of, and assumptions concerning, the curriculum have not been aired: notions, however, that are of fundamental importance in the process of curriculum construction.

First, we consider the model of the curriculum which is to be adopted.

The innovations of recent times have shown not only that many more mathematical topics can be taught at a school level than was previously thought possible, but also that the time at which topics are introduced can vary considerably. For example, matrices can be successfully taught to students at different age-levels, although, of course, the age of introduction should determine the methods used. Thus it has been demonstrated that there is no canonic (natural) order in which mathematical topics should be considered. This is not to say that one accepts the Brunerian thesis that 'every subject can be taught effectively and in an intellectually honest form to any child at any stage of development' (Bruner, 1960*b*). Clearly, teaching category theory to ten-year-olds would prove hard-going whatever form of presentation one adopted. Moreover as Freudenthal (1973*b*), among others, has pointed out, 'mathematics has seriously suffered under the falsifying tendencies in adaptations of mathematical subject matter to school level'. One must be careful when simplifying, not to distort aims, emphases or even mathematical truth.

Of course, there are some logically necessary sequences but these exercise comparatively few global constraints on curriculum building: the designer retains considerable freedom. Recently, there have been many attempts to determine the extent to which the curriculum should be determined by psychologically necessary sequences. Mathematical activities can then be planned bearing in mind a theory of the mental growth of the child. The authors of ATM (1977) are not alone in believing that such 'theories of mental growth often exert an anti-experimental, conservative influence on mathematics teaching'. Our present understanding of cognitive development would, indeed, hardly seem sufficiently secure to serve as a determinant of (rather than offering guide lines to) curriculum construction. Clearly, this is an area in which further research is vital.

Yet, notwithstanding the varying weights given to these considerations, with very few exceptions the curriculum projects have produced courses which are essentially linear. Indeed, the move from a fragmented arithmetic, algebra, geometry course to one demonstrating the 'unity' of mathematics has been accompanied by an increase in the structure provided within textbooks. It has become increasingly difficult for teachers to vary the order in which chapters are presented and in which, it is assumed, pupils will learn their mathematics. Elaborate flow-diagrams are often supplied to which materials conform. Yet is this a valid model of mathematical learning? Is learning mathematics analogous to climbing a tree (first a common trunk, then a variety of branches to be tackled in a directed manner), or to solving a monster jig-saw-puzzle (building isolated groups of pieces and then combining these by means of well-chosen links to form even bigger aggregates).

Some innovators (for example, members of the Leapfrogs group in England) believe that the latter analogy has something to offer and have attempted to design non-sequential activities consonant with a 'non-linear view of mathematics'. To date allegiance to 'linear' and 'non-linear' views can only be dependent upon personal beliefs: there is insufficient evidence either to disprove or to substantiate the claims of the 'non-linearists'.

Here, however, we have a revolutionary view of how one might design a mathematics curriculum and one which merits exploration in detail, as does, indeed, the more general problem of what models of the curriculum can act as an aid to curriculum design.

Exercises

1. Investigate the context in which Bruner's thesis (p. 249) was expressed. Barbel Inhelder had added: 'provided they [the subjects] are divorced from their mathematical expression and studied through materials that the child can handle himself'. What significant changes follow if we accept this proviso? Exemplify.
2. Give examples of mathematics which has been presented at school level in a 'falsified' manner.
3. Take sample flow-diagrams from books and project literature and consider in which instances the relative position of topics is firmly determined by either mathematical or psychological considerations.
4. In the 'jig-saw model' of learning we referred to combining groups by means of 'well-chosen links'. In a jig-saw puzzle the links are pre-determined. In our learning they are not. Discuss 'links' in relation to learning. Can they be supplied by the teacher? Must the student actually form them and, if so, how are we as teachers to encourage their formation?
5. Find and study other models which have been proposed for the curriculum.
6. 'Our so-called perfect textbooks rank amongst the greatest evils to be found in our present system of instruction. The very completeness and so-called strictly logical arrangement of these books, are the great causes which render them unsuitable for the development of the juvenile mind. The system which these books pursue is not the system which nature lays down for the development of the human faculties.' (Tate, 1857). Comment.

2.2. *Problems of choice in the curriculum*

The Leapfrogs group mentioned above argues (1975) that 'the best kind of learning is that in which the learner makes the subject matter . . . his own. And this is best done by providing a wide variety of significant and potentially interesting activities and stimuli from which the learner can make a choice'.

This brings us to a second point for consideration in curriculum design, namely the degree to which 'choice' is allowed, both on the part of the teacher and the student. As we have previously emphasised, even in the most centralised system the teacher has considerable opportunities in the classroom for exercising choice. Yet this freedom will fall short of, say, determining on what mathematics his pupils will be examined or the mode of examination to be used. In other systems, considerable choice will be permitted, even in the matter of external examinations. The advantages of permitting some degree of choice are not hard to find. It allows the teacher to vary the content of what he teaches from year to year and so helps counteract stagnation; new topics can enter the classroom in an experimental (and limited) fashion; syllabuses can be tailored to the individual interests and beliefs of teachers. Yet if a wide variety of choice is permitted at one level of education considerable problems will ensue for those designing curricula for the next, higher level. Thus in England there have recently been complaints from universities about the wide variety of school courses which their entrants have followed. Entrants, it is argued, possess comparatively little common knowledge: thus the university lecturer must begin correspondingly far back, knowing that whatever he chooses to teach he will be 'going over old ground' for many of his students. Calls have been made, therefore, for a 'core curriculum' common to all sixth-form A-level courses. Even allowing for the fact that a uniform curriculum does not ensure uniformity of competence and understanding, which is what the universities really want, the question still remains as to how big a 'common core' should be. What degree of choice should be available to the teacher in the way of options? Should a 'core' comprise 60%, 70%, 80% or 90% of the course? These are not easy questions to be answered: one fears that the answer will be determined more by the amount of material which higher education deems desirable, than by more general educational considerations. But if options are introduced, who does the opting: teachers or pupils? As we have said, there are strong arguments for allowing the teacher to select topics in which he has a particular interest. That, for motivational and other reasons, it is the pupil who should make the choice can also be argued (and is the principle enunciated by the Leapfrogs group). Giving choice to pupils raises not only the same problem as that described above concerning the core curriculum – what percentage of mathematical experiences do we expect to be shared? – but creates additional teaching and administrative problems. Once again a delicate compromise is called for; the same answer will not suffice for all teachers.

Exercises

1 Write an essay on the implications of 'choice' in the curriculum.
2 How feasible is it to offer not only a choice of content, but also a choice of learning styles? Discuss the way in which one might resolve the rival claims of (a) letting the teacher choose his 'preferred' teaching style; (b) accommodating to a pupil's 'preferred' learning style; (c) enriching the student's battery of effective learning styles.
3 Set out a case for allowing the teacher some choice in determining the examination syllabus his students follow.
4 'It is obvious that young students cannot be subjected to the vagaries of individual teachers. In this sense, the claim for freedom of the teaching profession is nonsense, but the general community is very incompetent to determine either the subject matter to be taught or the permissible divergencies to be allowed, or the individual competence. There can only be one appeal, and this to general professional opinion as exhibited in the practice of accredited institutions.' (Whitehead, 1933.) Discuss! (Is it equally 'true' if one substitutes 'old' for 'young' in the first line?)

2.3.   *Problems of differentiation*

Although 'choice' would seem to bring difficulties in its wake, it has been argued (see, for example, Sawyer, 1978) that 'choice' might prove to be the solution to one of today's principal curricular problems, the differentiation of students and courses.

As we have indicated (p. 22), many secondary-school systems were planned, or evolved, as two- or three-track systems. Often three distinct streams could be identified: academic, technical (vocational) and 'the rest'. Usually, students were selected for the first two streams by means of intelligence tests; although as the data quoted by Hitpass (p. 22) indicates, the selection was also to a considerable extent by social background. The mathematics courses provided within the three types of school differed substantially, as often did the methods of teaching. The Grammar School/Gymnasium/Lycée followed an academic curriculum founded on tradition. It was the most prestigious of the three types of school. The Technical School/Mittelschule/Collège Technique occupied a middle position. It often attracted neither the intellectually most able, nor great prestige, yet performed a useful, underrated service. The curriculum was vocationally oriented and frequently designed to fit the needs of local industry and commerce. Practical mathematics, such as technical drawing, was emphasised. The third, 'Secondary Modern' stream evolved gradually from the old elementary and primary schools as the age of compulsory education was raised. Looking back, one notes how during this process surprisingly little attention appeared to be paid to curricular aims. The general feeling appeared to be that an additional year or two of schooling would *per se* be beneficial: detailed consideration of what should be done in that time was superfluous. The result was that the mathematical curriculum for such schools was rarely thought out from scratch. It either took the form of an expanded course of arithmetic and mensuration, based on what had happened in the elementary school, or a watered-down version of what was done in the grammar or technical schools. As time went on efforts were made (see, for example, *Mathematics in Secondary Modern Schools,* Mathematical Association, 1959) to provide a specially devised curriculum suitable for the majority. However, such moves were overtaken in many countries by the change to comprehensive education. The social problems caused by dual and tri-partite educational systems were obvious to all. On the other hand, multilateral comprehensive schools, in which academic divisions were retained within a single school, found few allies. Why bring children together into the same school and then separate them again by setting and streaming? 'Mixed-ability' teaching became the vogue in several countries. Attempts were made to remedy social ills by social, administrative means. As a result educational aims and objectives have often been given insufficient emphasis. One course for all has almost invariably meant a watered-down academic course for all (even those capable of taking it neat). Differentiation into streams, when it has come, has often been too late for children of lower abilities, who still find themselves with inadequate arithmetical foundations. Those in the middle ability range find that employers have more to say about their basic shortcomings than their more esoteric achievements. Moreover, not only has uniform content been imposed on all, but the same teaching methods are used with no consideration of the fact that children of lower ability may well fail to respond to methods which can be successfully used with high-fliers, who in turn may be bored to tears by methods well-suited to the average student.

There has been a clear mismatch between pupils, methods and content. Yet how is this problem to be resolved? Sawyer (1978) suggests that schools should offer 'a free choice between mathematics in a practical setting, . . ., and a theoretical course in which pupils were encouraged to read ahead at their own rate.' The idea has its attractions (but what are the implications concerning 'reading ahead'?). Experience, however, shows that, when any such choice is given, it is the prestigious academic course which exerts the greater drawing power. Of course, it places the onus of choice on children and their parents, as in the old days it was placed on the administrators responsible for selection. Just how much responsibility, however, should teachers be called upon to shoulder?

The problem is a very complicated one involving many social and educational aspects. There will be no simple solution; but the devising of appropriate procedures and courses to alleviate current unease is a curricular priority for the coming decade.

Exercises

1 Investigate the attempts that have been made in your country to cope with the mathematical needs of the wider range of students entering secondary education.

2 In what ways can/do the learning styles of the less able pupils differ from those of the more able? How can we respond to these differences in our mathematics teaching?

3 Discuss the problems raised by the classification/differentiation of children.

4 Take Sawyer's suggestion above and attempt to devise appropriate practical and theoretical courses. What constraints will have to be satisfied?

5 'If there be any who still hold that without [mathematics] there is no secondary education, then they must face the distressing fact that nature has denied secondary education to a large part of normal humanity; for the evidence is conclusive that very many children, perhaps even a majority, are incapable of progressing any distance in [this subject] or of extracting any substantial benefit from [its] study.' *(Secondary Education,* Report of the Advisory Committee on Education in Scotland, 1947.*)* Reactionary nonsense or Scottish common-sense? What are the arguments for and against making mathematics a compulsory part of the school curriculum for all students under the school-leaving age?

2.4. *Problems of basic research*

We have already referred to our inability to ensure that students acquire even simple technical mastery. Indeed, if we are truthful, we must admit that we know little about what really happens in the learning process of either the individual or of a group in the classroom, or about the processes of interaction in the classroom, and that in many respects the teacher remains an unknown being (especially in so far as how his cognitions influence his actions). For too long such knowledge was regarded as self-evident and was not even questioned. It was forgotten that much of what we accept for knowledge in this domain are merely working hypotheses which may well help us to produce explanations, but which prove of limited value when we attempt to use them in the reverse sense, i.e. as principles from which procedural decisions can be deduced. Equally there are vast gaps in our understanding of the relationship between school-education and society. The demands which society imposes are almost always concrete ones. Yet, within the non-vocational school, attempts to meet these special requirements have to be generalised as more or less abstract goals. But we cannot teach goals: they once more have to be transformed into the concrete terms of content and material. So translation processes on different levels have to be carried out. Most of these are neither controlled nor coordinated: they occur, but are not planned. What is rarely questioned is whether these two processes of generalisation and of concretisation are in fact compatible. Yet it would seem the case that, say, the nous (commonsense, knowledge, resourcefulness, enterprise) required for book-keeping cannot be conveyed by simulating it in a school-version reduced to the level of children's understanding. The models of the world which we present to students are often too simplified. When problems are stripped down to their mathematical essentials, they no longer serve as general preparation for life outside the school, but merely become vehicles for exercises within mathematics itself.

Problems in the sociological domain do, moreover, differ in nature from those in cognition and psychology, in so far that the difficulties no longer lie in the intransigence of the matter itself. Now, the points at issue are plain to see. However, a major barrier is that every step can have a political implication: even the innovators' demands for elucidation of intricate relationships can provide cause for concern.

The short-comings which we have noted so far in this section cannot all be rectified by curriculum innovators alone. During the reform period it became a truism that progress in curriculum development depends upon interdisciplinary work and the unification of theory and practice. No longer are there projects with large staffs contributing from various specialisms. Yet the need for cooperation has not diminished. New models which provide for the drawing together of talents from different fields will have to be created. In the past, the difference in social reputation between schools and universities has caused prejudices and handicapped cooperation in many countries. It is obvious, however, that dialogues can only be successful when each in his own domain acquires and recognises the level of competency that the other possesses in his. Cooperation must be based on the acknowledgement of mutual dependency and respect.

In particular, only through the joint efforts of practitioners and researchers are we likely to gain that essential understanding of the classroom, an understanding which so far eludes us and which is still the most significant gap in our knowledge. For all those involved in curriculum development it is still a most startling fact that our knowledge of the classroom remains based on haphazard experiences, on personal observation or the impressions of others: beyond that we have little more than speculations and assumptions.

Few of the projects of the 1960s and 1970s made any attempt to base their approaches upon educational research specifically initiated to solve curricular problems. The reasons are not hard to find: Goodlad (1969) summed up the position when he spoke of 'the paucity of ordered "findings" from curriculum research – findings in the sense either of scientific conclusions from cumulative inquiry or of tested guidelines for curriculum decisions'. Some projects, as we have seen, chose to draw on more general findings within educational research. In general, however, research results tended to be used more as crutches to support decisions taken on philosophical grounds, or as cudgels with which to attack the opposition, rather than as frameworks for curriculum design or determinants in decision-making.

The paucity of studies in curriculum research was, not unnaturally, matched with a paucity of models that could be used for studies within curriculum research.

Those models imported from science and founded on the experimental verification of hypotheses were hardly suited to meeting the specific demands of curricular research. Isolation of variables is almost impossible, both in theory and in practice; the time-span required for observation and experiment can be many years; the selection of samples of adequate size and randomness can prove extremely difficult. Changes occur within the sample and in the experiment's educational setting which cast doubt upon the validity and usefulness of the results.

Criticisms such as 'This study represents a very thorough, statistically sophisticated analysis of fallible data' (Aiken) and 'Alas for these hypotheses, most may belong to an era now spent, regarding both the status of mathematics education in the United States and the relative status of the sexes' (Nibbelink) illustrate the problems faced by the researcher. Comments such as 'the overwhelming amount of data generated by this study and the complexity of the statistical models employed made it even more difficult to make sense out of the results' (Lester) and 'the reporting of the data is so inadequate that interpretation by the reader is almost impossible' (Fennema) illustrate the problems so frequently encountered by those who would wish to make use of the research. (All the above quotations are taken from *Investigations in Mathematics Education*, **8** (3), 1975, a special issue devoted to critical analyses of the NLSMA reports (see chapter 7).)

Carrying out valuable research is one thing (which may have its personal rewards in terms of postgraduate degrees or academic advancement), presenting the results in a form in which they can be read and understood by the community of mathematical educators has too frequently been thought of as another. It is essential if research is to be looked upon as an aid to teachers and curriculum designers, rather than merely a way of creating employment for academics, that due weight be given to problems of communication.

As we have indicated, however, research in the field of mathematics education poses problems on a totally different scale from those encountered within mathematics or pure science. In the latter disciplines useful research work can be undertaken immediately after a first degree course. The work can often be completed quickly, communicated rapidly and assimilated into that growing body of accepted findings. Any criticisms concerning methodology and validity can fairly be made after reading a relatively short research paper. In mathematics education not only a sound knowledge of mathematics is required, but also training in educational research techniques and that understanding of education which only years of experience can bring. The problems of carrying out research are, as we have seen, considerably more difficult, as are those of communication and, most importantly if mathematics education is to emerge as a 'discipline', those of criticism and assimilation. Moreover, in only a handful of centres throughout the world does one have the amount of research activity and the number of research students which are accepted as a matter of course in countless mathematics and science departments. The resulting fragmentation in research funds and effort, and the isolation of researchers has a detrimental and limiting effect on what can be accomplished. There would appear to be an obvious need for greater coordination and/or concentration of research interest.

Some interesting possibilities for new research orientations in the field of the curriculum are to be found in Walker (1973). Indeed, already one can detect a growth in the use of 'non-standard' research models (cf. p. 186–7). Some of these studies, based on classroom observation, would seem to be more successful in casting doubt on traditional, statistically-based research procedures, than in offering guidelines of the type which Goodlad (see above) sought. The studies reveal all too clearly the complexity of the classroom. It is right that this complexity should be emphasised and simplistic attempts to conceal or forget it criticised. However, there is a limit to the value of this work: having accepted the complexity one must devise models and procedures that will enable researchers to produce findings of positive value to curriculum designers. Knowing what does not work is valuable, but knowing what *does* is of immeasurably more use! Yet we can never expect to know what will work, in a readily replicable way in all classrooms, with all teachers and all pupils. We must be satisfied with lower level guides. Here again some help has been forthcoming and one hopes more will come through detailed classroom investigation and observation focused on the way in which small units of the curriculum can be presented to and learned by students. Meticulous observation and analysis of the way in which misunderstandings occur, of blockages and aids to learning, of the interaction between mathematical and other learning, can provide information of great value to those who must design curricula.

Exercises

1 What is meant by 'mastery' in mathematics?
2 'In many respects the teacher remains an unknown being.' Expand (or refute).
3 'The models of the world which we present to students are often too simplified.' Exemplify (or refute) and discuss consequences and possible alternatives.
4 'Each in his own domain (must acquire and recognise) the level of competency that the other possesses in his.' Expand and explain.
5 Discuss the problem of communicating research results. Take specimen research publications and ask for whom they appear to have

been written. In what ways is the intended audience influential? Study and comment upon the popular press 'reviews' of research findings.

6   Why are the problems of 'criticism and assimilation' central to the emergence of mathematical education as a 'discipline'?

3.   The teacher and curriculum development

Even in Britain only a small proportion of teachers consciously participate in curriculum development. In other countries, where the educational system does not encourage such participation, the proportion is still smaller. Yet every teacher is invoved in curriculum development, whatever curriculum he follows, and there are obvious reasons why he should know as much as possible about its construction and be able to examine it critically.

When dipping into old textbooks one is often astonished to find that they ran almost unchanged through 20 or 30 editions. Teachers may well have used them as children and then throughout their teaching career. As a result, their teaching was based on a well-proven, conventional understanding of content and methods. Those times and curricula have gone. Nowadays, the teacher is often forced to rely on his own understanding and powers of interpretation. No wonder then that he will often welcome not only classroom materials and operating instructions, but also additional information concerning premises and intentions. Yet how often does such information serve more than publicity purposes? And how frequently do projects and textbook writers succeed in adhering to their published principles? The teacher's position is aggravated by the quantity of competing material thrust before him: an enormous amount has been produced in a very short time. Moreover, materials which although outdated still offer the teacher considerable security remain on the market. The attractions of such traditional materials have increased as in some countries innovation has come to a somewhat abrupt end, and reaction set in, often resulting in the setting aside of both the errors and successes of the reforms: in truth, 'the good is often interred with their bones'. Curiously enough, at a time when many were puzzled by the vast effort devoted to, and expenditure on, curriculum development, few seemed to have foreseen that this epoch would end so quickly and would leave us with so many unanswered questions.

The teacher whose educational system allows him some freedom in determining the curriculum he will follow faces a far from easy task. For him a knowledge of curriculum development is essential if he is to make a responsible choice. Yet similar considerations will also apply to the teacher who has a curriculum imposed upon him. If he wishes to assume the role he *can* play in his students' learning, then he will need to be aware of the weaknesses in the materials he is using and of how to compensate, and of how to take advantage of their strengths. He, too, will need competency in the domain of curriculum development.

A teacher must understand the materials he uses, and not only in the sense of mastering the mathematics they contain. Not every teacher will be able to analyse the textbook he uses with respect to its underlying intentions and the adopted means; how these work together and conform to his own intentions and experience. Yet this ability vitally affects the outcome of his work. One clear way in which an understanding of materials can be arrived at is by comparing them with others. As our case studies have shown, existing materials exhibit a wide range of goals, content and methods. Considering the place which any particular materials occupy in the spectrum of possibilities will help us the more easily to recognise their salient features. Here, knowledge of curriculum development proves most valuable.

Being aware of a variety of different approaches and of the materials representing them is, of course, useful in other respects. Not only do they help us to see familiar materials in a new light, they may well provide inspiration and suggest more appropriate responses to particular classroom situations. Clearly, it would be advantageous if every teacher had access to a variety of texts and project materials: and had time provided to study them!

Such a relationship between the teacher and teaching materials would, however, mean a considerable change in current practice. Indeed, in general it would mean an important change in the teacher's professional self-consciousness. A critical attitude towards teaching materials can only arise from a critical understanding of one's role as a self-reliant director/manager/guide of one's students' learning. It must be based on professional competency, comprising sound mathematical knowledge as well as a knowledge of pedagogy in its widest sense. In two words, the teacher here described would be competent and autonomous.

Is asking this to ask for the moon?

First it must be emphasised that such aptitudes and abilities cannot be conceived of in terms of abstract knowledge and skills to be provided by means of a well-designed and thorough training. If knowledge is one side of the teacher's competency and autonomy, then his experience is the other. The ability, and willingness, to decide and to act responsibly in practice depends on both knowledge and experience. There are few who would suggest that a new recruit to teaching, fresh from a university or college of education, is ready to exercise autonomy or is as fitted to do so as is his older colleague.

Yet this does not mean that the growth of competency and autonomy should be left to haphazard, and more or less automatic, processes inherent in the teacher's individual development. It is obvious that this growth should be highly dependent upon outside conditions too, the most important being the teacher's training and the environment in which he works. It is, of course, a delicate matter to call for autonomy in an educational system or society which does not concede it. We are all aware that there are those powerful individuals within education who have grasped and wielded autonomy under all sorts of social conditions and régimes. Nevertheless, it has to be stressed that autonomy must be granted to all those who demand, and can exercise, it.

When considering that essential foundation of 'sound mathematical knowledge' one's thoughts turn instinctively to the teacher's pre-service training. It must, however, be accepted that no matter how well planned that may be, it cannot provide the background mathematical knowledge sufficient to span a teaching career of forty years. Further study will be needed. Recently attempts to improve the academic standard of teacher-training have not been universally successful: 'partially-digested courses on branches of mathematics beyond a student's capacity and interest only make him think that misunderstanding is the norm in mathematics learning. He then passes this view of mathematics on in his own classroom.' (Royal Society, 1976.) Even when it has been realised that mathematical *confidence* should be the *sine qua non* for mathematics teachers, we have not always been able to provide it in our students. The need to prepare student-teachers for further study, as indicated by Moseley in 1846 (p. 28), has been given considerable attention in recent years and there has been a welcome growth in the use of project work, assignments, etc.

Balancing the rival claims of 'general education' and 'professional training' will, however, remain a problem for all institutes supplying pre-service education. Further attention will also have to be paid to delineating more precisely that mathematics which we should like teachers of mathematics (at various levels) to know, and that of which we wish them to be aware.

Improved pre-service education cannot, however, solve all our problems. There will also be a need for planned in-service education. This is now particularly vital as the educational systems of the Western world contract with a consequent fall in the number of new entrants to teaching. The vast majority of those who will be teaching in A.D. 2000 are already in post.

As we have seen earlier, the amount and varieties of in-service education provided have greatly increased in recent years. Nevertheless, in most countries it remains uncoordinated and haphazard in nature. It is a right (often grudgingly or infrequently bestowed), rather than an obligation for which due allowance is made. The James Report on *Teacher Education and Training* (London, HMSO, 1972) placed great emphasis on the need to provide a wide range of activities which would enable teachers to continue their personal education and extend their professional competence:

> [These would cover] a wide spectrum, at one end of which are evening meetings and discussions, weekend conferences and other short-term activities, with limited and specific objectives and taking place usually but not always, in the teachers' own time. At the other end are long courses leading to higher degrees or advanced qualifications, and requiring the release of teachers for full-time attendance at suitable establishments. At this end of the spectrum, too, may be periods of release to take part in curriculum development and evaluation, or in other projects and investigations. For some teachers there may be periods of secondments to fields outside teaching, so that they may widen their experience and thereby enrich their contribution to the schools. Between lies a wide variety of courses and other activities, of different lengths and patterns, serving many different purposes. These activities may be part-time or full-time, or may include periods of both. They may take place entirely during school hours and require the release of the teachers concerned, or be entirely in the teachers' own time, or they may involve a mixture of both. For this large and complex field, it is clear that 'inservice training', however convenient as shorthand, is a very misleading term.

The report went on to recommend that 'as soon as possible' teachers should be entitled to release with full pay for in-service education on a scale of one term in five years.

The recommendations of the report were accepted, the 'principle of unripe time' invoked, and the report left to gather dust until economic circumstances improved. Yet even the acceptance of the recommendations was a step forward: regrettably the need has not even been realised in some countries.

Yet the James Committee (like those committees which reported on the raising of the school-leaving age, see p. 171) was unable to provide flesh for the skeletal recommendation that there should be increased opportunities for in-service education. Even after twenty years of experiments and increased opportunities there would still appear to be considerable doubt about the most effective agencies for providing in-service education and the most useful forms that it can take. Some small-scale studies have been carried out in an attempt to find out 'what works', but much more remains to be discovered and acted upon.

Yet another complicating factor concerning the provision of in-service education springs from the fact that not all teachers are convinced of the need for it. Freedom to innovate also implies freedom not to: the educational system will contain autonomous conservators in addition to innovators! There are, of course, as many reasons for opting out of innovation as for opting in. Sheer idleness is to be deplored; but the teacher who has taken pains to learn about an innovation, to study materials, to visit classrooms in which the new curriculum is being followed, and who then decides that on balance it is not for him or his students is not acting unprofessionally. Far better that than become a crowd-follower who, often with the best of intentions, finds himself in a new situation with which he is unable to cope.

Autonomy is not best served if we merely attempt to jump on every bandwagon that passes. Yet it is our professional duty to acquaint ourselves with new ideas and to make arrangements to see new methods in action and, where necessary, to make small-scale experiments. Just as a doctor will keep himself informed about new medicines and techniques, or a solicitor about changes in the law, so must teachers keep abreast of developments. This must be seen as part of a teacher's professional responsibility. As we have indicated, the assumption of such responsibility must be balanced by the provision of time and facilities to enable teachers to study and to plan; and their status in the community, both financial and social, must be related to the responsibilities they bear. This will not be accomplished overnight, and will certainly necessitate a detailed study of incentives, accreditation schemes, etc.

The key to successful in-service education is, however, the attitude generated within individual schools. Here, the headteacher has a major role to play as does, in those countries where such a post exists, the head, or chairman, of the mathematics department. He will be responsible for keeping the curriculum of the school under constant review and for guiding the work and the careers of other mathematics teachers in his schools. Here not only mathematical and pedagogical skills will be needed, but also managerial ones. These are vital positions within schools and ones for which improved training facilities would seem essential. Here educators may well have something to learn from industry and the armed services, where training for management roles is treated far more seriously.

Training for participation in curriculum development, therefore, must be planned and continuous. The new entrant to teaching, depending on his qualifications and aspirations, will initially take up a position on the spectrum ranging from the textbook slave to the individualist designing and assessing his own work. Ideally, as a result of involvement in local work, at a teachers' centre or elsewhere, and through guidance provided by his head of department, he will accept additional responsibilities and with growing professional competence, both pedagogical and mathematical, be enabled to shift his position towards that of the full professional. Such a planned progression would seem essential if further curriculum development is to be successful. For this must be founded on, and integrated with, an effective in-service programme.

It is a characteristic feature of the English school-system (on which many of the above observations are based) that in-service education is often identified with out-of-school activities concentrated upon teachers' centres. In this way in-service education becomes institutionally connected to curriculum development, a relationship which offers much to both sides. It is this which makes the English teachers' centre a most valuable innovation, and one which has still to reveal its full potential. We have already noted how, for the teacher, developmental work in all its complexity can prove the most appropriate and stimulating base on which to found in-service education. The centres can also, of course, introduce the teacher to developmental work at a variety of levels, from loose acquaintance, through small-scale work (e.g. the joint preparation of single units for classroom use), up to participation in the work of major projects. In such a manner, a majority of teachers can advance in professional consciousness, autonomy and engagement.

On the other hand, the gains for curriculum development will also be immense. Here we are not insisting or even suggesting that all materials should be personally designed by the teacher concerned, or even that the teacher should have played an active part in their design. Curriculum development will have to coexist with 'non-participant' teachers: but the latter will still have a vital contribution to make providing they are familiar with the significance and the workings of innovation. The lessons of the reform period we have recently witnessed are that most attempts to enforce radical changes in practice have been subject to trouble and distortion and that only rarely have original intentions been realised. If innovation is to proceed more satisfactorily in future then it is essential that we ensure better understanding and acceptance by teachers. Consequently, one of the most significant tasks for future work in the field of curriculum development is to broaden the base of innovation. This applies even to changes introduced some time ago: misunderstanding is still rife and occasionally gives rise to surprising mismatches of intention and outcome. But a basic acceptance of, and engagement in, innovation is an even greater goal. In 'official' circles, there is now little enthusiasm for curriculum development: its stock stands low. As a result, many large-scale initiatives have ended and there has been no support forthcoming for further work. At such times, as the history of education exemplifies, it is within the schools that ideas are preserved and claims maintained: there they are incorporated into the experience of teachers. True, there are many teachers who have been disappointed by the failures and errors of recent curricular changes. We must not, however, fear that enthusiasm for improving school education will die out. Alas, it has been demonstrated that innovation and improvement are not necessarily synonymous. Nevertheless, it is only through curriculum development that our goals for school education will be realised.

Appendix 1

A Comparative Guide to School Structures

Age (yr)	Grade	England[a]	Federal Germany	USA[b]
5–6	K		Pre-school	
6–7	1	First		(Primary)
7–8	2		Primary	Elementary
8–9	3			
9–10	4	Middle		(Intermediate)
10–11	5		(Hamburg-Berlin) — Orientation level — Hauptschule, Realschule / Gymnasium — Comprehensive School	
11–12	6			
12–13	7			Middle/Junior High School
13–14	8	Junior Secondary	Secondary I	
14–15	9			
15–16	10		For Hauptschule	(Senior)
16–17	11	} Sixth-form	} Secondary II	} High School
17–18	12			

═══ indicates years of compulsory education

[a] The pattern illustrated is only one of a variety of structures to be found. Alternatives include 5–7, 7–11, 11–18; 5–9, 9–13, 13–18. In England the break between primary and secondary education traditionally occurred at 11 yr. The introduction of middle schools has tended to delay this break. In the USA the break is now tending to occur earlier, at 11 rather than 12 yr.

[b] Two common US patterns are shown; others combine elementary and middle schools in an 8-year elementary school or combine junior and senior high schools in a 6-year secondary school.

Appendix 2

A Glossary of Abbreviations

The reader is referred to the various reports of the International Clearinghouse on Science and Mathematics Curricular Developments published by the Science Teaching Center, University of Maryland, for further details concerning projects.

AAAS   American Association for the Advancement of Science (USA)
AIGT   Association for the Improvement of Geometrical Teaching (UK)
ATM   Association of Teachers of Mathematics (UK)

BCMI   Boston College Mathematics Institute (USA)

CAI   Computer Assisted Instruction
CBMS   Conference Board of the Mathematical Sciences (USA)
CIEM   Commission Internationale de l'Enseignement Mathématique
CMP   Caribbean Mathematics Project (West Indies)
        Continuing Mathematics Project (UK)
CSE   Certificate of Secondary Education (UK)
CSMP   Comprehensive School Mathematics Program (USA)
CUPM   Committee on the Undergraduate Program in Mathematics (USA)

DIME   Development of Ideas in Mathematical Education (UK)

GCMP   Greater Cleveland Mathematics Program of the Educational Research Council of America (USA)

ICMI   International Commission on Mathematical Instruction
IGE   Individually Guided Education (USA)
IMU   Individualiserad Matematikundervisning (Individualised Mathematics Instruction) (Sweden)
IOWO   Instituut voor Ontwikkeling van hat Wiskunde Onderwijs (Institute for Development of Mathematical Education) (Netherlands)
IPI   Individually Prescribed Instruction (USA)

IREM	Institut de Recherche pour L'Enseignement des Mathématiques (Institute for Research in Mathematics Teaching) (France)
KMK	Ständige Konferenz der Kultusminister (Standing Conference of the Ministers of Culture) (Federal Germany)
MA	Mathematical Association (UK)
MAA	Mathematical Association of America (USA)
MINNE- MAST	Minnesota Mathematics and Science Teaching Project (USA)
MMCP	Mathematics for the Majority Continuation Project (UK)
MMP	Mathematics for the Majority Project (UK)
NCTM	National Council of Teachers of Mathematics (USA)
NIE	National Institute of Education (USA)
NLSMA	National Longitudinal Study of Mathematical Abilities (USA)
NSF	National Science Foundation (USA)
OEEC	Organisation for European Economic Co-operation
SED	Scottish Education Department
SEED	Special Elementary Education for the Disadvantaged (USA)
SMG	Scottish Mathematics Group (UK)
SMILE	Secondary Mathematics Individualised Learning Experiment (UK)
SMP	School Mathematics Project (UK)
SMSG	School Mathematics Study Group (USA)
SSMCIS	Secondary School Mathematics Curriculum Improvement Study (USA)
UICSM	University of Illinois Committee on School Mathematics (USA)
UNESCO	United Nations Education, Scientific and Cultural Organisation
UMMaP	University of Maryland Mathematics Project (USA)
USMES	Unified Science and Mathematics for Elementary Schools (USA)

Bibliography

In addition to the publications referred to in the text, we list some other papers and books which readers are likely to find of interest and help.

Reports and Books published by Ministries, Bodies, etc.

(i) International organisations

OECD/CERI (Paris)
  Case Studies of Educational Innovation
    I *At the Central Level* (1973)
    II *At the Regional Level* (1973)
    III *At the School Level* (1973)
    IV *Strategies for Innovation in Education* (1973)
  Innovation in Education: Sweden (1971), *United States* (1971), *Germany* (1971), *England* (1971).
OEEC (Paris)
  New Thinking in School Mathematics, 1961.
  Synopses for Modern Secondary School Mathematics, 1961.
UNESCO (Paris)
  New Trends In Mathematics Teaching **1** (1966), **2** (1970), **3** (1972), **4** (1979).
  Interactions between Linguistics and Mathematical Education, 1975.
UNESCO (Hamburg)
  Curriculum-Entwicklung in der Bundesrepublik Deutschland, 1975.

(ii) National bodies and associations

Association of Teachers of Mathematics (UK).
  Notes on Mathematics for Children, Cambridge University Press, 1977.

Board/Ministry of Education, Department of Education and Science (UK).
  Report of the Schools Inquiry Commission 1868 (Taunton Commission), HMSO, 1868.
  The Education of the Adolescent (Hadow Report), HMSO, 1926.
  Secondary Education with Special Reference to Grammar Schools and Technical High Schools (Spens Report), HMSO, 1938.

Curriculum and Examinations in Secondary Schools (Norwood Report), HMSO, 1943.
15 to 18 (Crowther Report), HMSO, 1959.
Secondary School Examinations other than the GCE (Beloe Report), HMSO, 1960.
Half our Future (Newsom Report), HMSO, 1963.
Children and Their Primary Schools (Plowden Report), HMSO, 1967.
Enquiry into the Flow of Candidates in Science and Technology into Higher Education (Dainton Report), HMSO, 1968.
Teacher Education and Training (James Report), HMSO, 1972.
Cambridge Conference (USA)
Goals for School Mathematics, Houghton Mifflin, 1963.
Goals for Mathematical Education of Elementary School Teachers, Houghton Mifflin, 1967.
Goals for the Correlation of Elementary Science and Mathematics, Houghton Mifflin, 1969.
Conference Board of the Mathematical Sciences (USA)
General Principles of International Collaboration in Mathematical Education, 1970.
Overview and Analysis of School Mathematics, Grades K–12 (NACOME Report), 1975.
IOWO (Netherlands)
Curriculum Development – A Strategy, *Ed. Stud. Math.* (1976). **7**, 351–62.
Mathematical Association (UK)
Mathematics in Secondary Modern Schools, Bell, 1959.
Why, What and How? Some basic questions for mathematics teaching, 1976.
Evaluation, 1979.
National Committee on Mathematical Requirements (USA)
The Reorganization of Mathematics in Secondary Education, MAA, 1923.
National Council of Teachers of Mathematics (USA)
A History of Mathematical Education in the United States and Canada (32nd Yearbook) 1970.
National Science Foundation (USA)
Panel Evaluation of 19 Pre-College Curriculum Development Projects. Dec. 8–12, 1975, 1976.
National Society for the Study of Education (USA)
Mathematics Education (69th Yearbook, Part 1), University of Chicago Press, 1970.
Open University (UK)
Styles of Curriculum Development (E203, Unit 7), 1976.
Royal Society (UK)
The Training and Professional Life of Teachers of Mathematics, 1976.
Schools Council (UK)
Evaluation in Curriculum Development: Twelve Case Studies, MacMillan Educational, 1973.
What's going on in Primary Mathematics, 1974.
Scottish Education Department
Secondary Education (Hamilton Fyfe Report), HMSO, 1947.

(iii) Projects

Details of textbooks are not given here. They can be found in Reports of the International Clearinghouse on Science and Mathematics Curricular Developments, Ed. J. D. Lockard, University of Maryland, 1963 onwards.

Leapfrogs Group, *Publicity Handout,* Hutchinson, 1975.
SMP, *Manipulative Skills in School Mathematics,* Westfield College, 1974.

SSMCIS, *Information Bulletin No. 4,* Teachers College, Columbia University, November 1969.
Information Bulletin No. 7, Teachers College, Columbia University, Spring 1973.

Books and papers written by individuals

Ahlfors, L. V. *et al.* (1962). On the mathematics curriculum of the high school. *Maths Teacher* **55,** 191–4; *Amer. Math. Monthly* **69,** 189–93.
Anderson, S. B., Ball, S., Murphy, R. T. & associates (1975). *Encyclopedia of Educational Evaluation.* Jossey–Bass. San Francisco.
Anderson, V. E. (1965). *Principles and Procedures of Curriculum Improvement.* Ronald Press Co., New York.
Ashlock, R. B. & Herman, W. L. (eds) (1970). *Current Research in Elementary School Mathematics.* Macmillan.
Austin, J. L. & Howson, A. G. (1979). Language and Mathematical Education. *Ed. Stud. Math.,* **10,** 161–97.
Ausubel, D. P. (1961). Learning by Discovery: Rationale and Mystique. *Bull. Nat. Assn. Sec. School Principals,* **45,** 18–58.
Ausubel, D. P. (1964). Some psychological and educational limitations of learning by discovery. *Arith. Teacher,* **11,** 290–302.
Avital, S. M. & Shettleworth, S. J. (1968). *Objectives for Mathematics Learning.* Ontario Inst. for Studies in Edn.
Barszcz, E. L. (1975). *Teacher Interviews, Second Grade,* CSMP Evaluation Report Series, No. 2–C–2. CEMREL, St. Louis.
Beberman, M. (1958). *An Emerging Program of Secondary School Mathematics.* Harvard University Press.
Beberman, M. & Vaughan, H. E. (1964). *High School Mathematics,* Heath. Lexington, Massachusetts.
Becher, R. A. & Maclure, S. (1978). *The Politics of Curriculum Change.* Hutchinson.
Beeby, C. E. (1970). Curriculum planning. In *Developing a New Curriculum,* ed. A. G. Howson. Heinemann Educational.
Begle, E. G. (1968). SMSG: the first decade. *Math. Teacher,* **61,** 239–45.
Begle, E. G. (1971). Research and Evaluation in Mathematics Education. *Report of Conference on Responsibilities for School Mathematics in the 70's,* SMSG, Stanford.
Begle, E. G. (1974). Review of *Why Johnny Can't Add. Nat. El. Principal,* **53,** 26–31.
Begle, E. G. (1979). *Critical Variables in Mathematics Education.* MAA and NCTM, Wiley.
Begle, E. G. & Wilson, J. W. (1970). Evaluation of Mathematics Programs. In Begle, E. G. (ed.), *Mathematics Education,* 69th Yearbook of the NSSE (Part 1), University of Chicago Press.
Bell, A. W. (1975). Curriculum Development in the Lyon Area of France. *Maths Teaching,* **72,** 46–7.
Bernstein, B. (1975). *Class, Codes and Control* (vol. 3), Routledge & Kegan Paul.
Bloom, B. S. *et al.* (1956). *Taxonomy of Educational Objectives: 1. Cognitive Domain.* Longman.
Bloom, B. S. *et al.* (1964). *Taxonomy of Educational Objectives: 2. Affective Domain.* Longman.
Bloom, B. S. *et al.* (1971). *Handbook on Formative and Summative Evaluation of Student Learning.* McGraw–Hill.
Bobbitt, F. (1918). *The Curriculum.* Houghton Mifflin. Boston, Massachusetts.

Bobbitt, F. (1924). *How to Make a Curriculum.* Houghton Mifflin. Boston, Massachusetts.

Branford, B. (1908). *A Study of Mathematical Education.* Oxford University Press.

Braunfeld, P., Kaufman, B. & Haig, V. (1973). Mathematics Education: A Humanist Viewpoint. *Educ. Technol.* **13** (11), 43–9.

Breny, H. (ed.) (1976). *The Teaching of Statistics in Schools.* Int. Statistical Inst., Voorburg, Netherlands.

Bruner, J. S. (1960*a*). On learning mathematics. *Math. Teacher,* **53**, 610–19.

Bruner, J. S. (1960*b*). *The Process of Education.* Harvard University Press.

Bruner, J. S. (1964). Some theorems of instruction illustrated with reference to mathematics. In *Theories of Learning and Instruction.* NSSE (63rd Yearbook).

Bruner, J. S. (1966). *Towards a Theory of Instruction.* Harvard University Press.

Bruner, J. S., Goodnow, J. J. & Austin, G. A. (1956). *A Study of Thinking.* Wiley.

Cane, B. (1973). Meeting teacher's needs. In *In-Service Training,* R. Watkins, Ward Lock.

Cane, B. & Schröder, C. (1970). *The Teacher and Research.* NFER.

Carson, G. St. L. (1913). *Mathematical Education.* Ginn & Co.

Chin, R. & Benne, K. D. (1969). General Strategies for Effecting Changes in Human Systems. In *Planning of Change,* ed. W. G. Bennis *et al.* 2nd edn. Holt, Rinehart & Winston.

Christiansen, B. (1976) Europe – past and present: an outline of major developments in the period 1960–1976 *and* European mathematics education – the future. In *Proc. AAMT Biennial Conference,* Perth, Australia.

Colburn, W. (1821). *Intellectual Arithmetic, Upon the Inductive Method of Instruction.* Reynolds & Co., Boston.

Compayré, G. (1908). *Herbert Spencer and Scientific Education.* Harrap.

Coombs, P. H. (1968). *The World Educational Crisis.* Oxford University Press.

Crawford, D. H. (ed.) (1975). *The Fife Mathematics Project,* Oxford University Press.

Cundy, H. M. (1976). *Caribbean Mathematics Project: An Evaluation Study.* British Council,London.

Cundy, H. M. (1977). *The Caribbean Mathematics Project: Training the Teacher as the Agent of Reform.* Experiments and Innovations in Education, No. 32, Unesco, Paris.

Curtis, S. J. & Boultwood, M. E. A. (1965). *A Short History of Educational Ideas.* 4th edn. University Tutorial Press.

Dahllöf, U. (1974). Curriculum Development in Sweden. In *Curriculum Development,* ed. P. H. Taylor & M. Johnson. NFER.

Dalin, P. (1973). *Case Studies of Educational Innovation: IV Strategies for Innovation in Education.* OECD.

Dalin, P. (1978). *Limits to Educational Change.* Macmillan.

Damerow, P. *et al.* (1974). *Elementarmathematik: Lernen für die Praxis?* Klett; Stuttgart.

Damerow, P. (1977). *Die Reform des Mathematikunterrichts in der Sekundarstufe I.,* Klett; Stuttgart.

Davis, R. B. (1964). The Madison Project's Approach to Theory of Instruction. *J. Res. Sc. Teaching,* **2**, 146–62.

Davis, R. B. (1965). *A Modern Mathematics Program as it Pertains to the Inter-relationships of Mathematical Content, Teaching Methods and Classroom Atmosphere (The Madison Project).* Coop. Research Project No. D–093, Syracuse University and Webster College, St. Louis.

Davis, R. B. (1967). *Final Report of Research Project No. D–093,* Syracuse University and Webster College, St. Louis.

Davis, R. B. (1967). *The Changing Curriculum: Mathematics.* Assn. for Supervision and Curr. Dev: Washington.

Davis, R. B., Jockusch, E. & McKnight, C. (1978). Cognitive Processes in Learning Algebra, *J. Child. Math. Beh.* **2**, 10–320.

Dee, J. (1570). *The Mathematical Praeface,* Watson, Reprint 1975.

DeMott, B. (1964). The Math Wars. In *New Curricula,* ed. R. W. Heath. Harper & Row.

Devaney, K. & Thorn, L. (1974). *Curriculum Development in Elementary Mathematics: 9 Programs.* Far West Laboratory, San Francisco.

Dewey, J. (1963). *Experience and Education.* Collier Macmillan.

Dienes, Z. P. (1963). *An Experimental Study of Mathematics Learning,* Hutchinson.

Dienes, Z. P. (1965). *The Power of Mathematics.* Hutchinson.

Dienes, Z. P. (1966). *Mathematics in Primary Education,* UNESCO.

Dienes, Z. P. (1973a). *Mathematics through the senses.* NFER.

Dienes, Z. P. (1973b). *The Six Stages in the Process of Learning Mathematics,* NFER. (English translation of 1970 French original.)

Dienes, Z. P. & Golding, E. W. (1967). *Geometry through Transformations.* Vol. 3: *Coordinates and Groups.* ESA.

Dieudonné, J. (1973). Should we teach modern mathematics? *Amer. Scientist,* **61**, 16–19.

Dilworth, R. P. & Warren, L. R. (1973). *Mathematics Improvement Programs: Final Report, Specialized Teacher Project, 1971–72.* Report submitted by the State Board of Education to the California Legislature.

Dworkin, M. S. (ed.) (1959). *Dewey on Education,* Columbia University.

Easton, J. B. (1966). A Tudor Euclid. *Scripta Math.* **27**, 339–55.

Eisenberg, T. A. (1975). Behaviourism: the bane of school mathematics. *International Journal of Mathematical Education in Science and Technology.* **6**, 163–171.

Eisner, E. W. (1967). Franklin Bobbitt and the 'Science' of Curriculum Making. *School Review,* **75**, 29–47.

Eisner, E. W. & Vallance, E. (eds.) (1974). *Conflicting conceptions of curriculum.* McCutchan.

Erlwanger, S. H. (1973). Benny's Conception of Rules and Answers in IPI Mathematics. *J. Children's Math. Beh.* **1** (2), 7–26.

Erlwanger, S. H. (1974). Case Studies of Children's Conceptions of Mathematics, Unpublished doctoral dissertation, University of Illinois, Urbana.

Fehr, H. F. (1966). SSMCIS, *American Math. Monthly,* **73**, 533.

Fehr, H. F. (1974). SSMCIS: A unified Mathematics Program. *Math. Teacher,* **67**, 25–33.

Fey, J. T. (1978). Change in mathematics education since the late 1950's – USA. *Ed. Stud. Maths.* **9**, 339–53.

Fey, J. T. (1979). Mathematics Today: perspectives from three national surveys. *Math. Teacher,* **72**, 490–504.

Flanagan, J. C., Shanner, W. M. & Mager, R. F. (1971). *Behavioral Objectives: A Guide to Individualizing Learning. Mathematics.* Westinghouse Learning Press, Palo Alto.

Ford, G. W. & Pugno, L. (eds.) (1965). *The Structure of Knowledge and the Curriculum.* Rand McNally.

Foshay, A. W. (1962). Discipline-centred curriculum. In *Curriculum crossroads,* ed. A. H. Passow. Columbia University, New York.

Freudenthal, H. (1973a). *Mathematics as an Educational Task.* Reidel.

Freudenthal, H. (1973b). What groups mean in mathematics and what they should mean in mathematical education. In *Developments in Mathematical Education,* ed. A. G. Howson. Cambridge University Press.

Freudenthal, H. (1978). *Weeding and Sowing.* Reidel.

Gagné, R. M. (1962). The acquisition of knowledge. *Psych. Rev.* **69**, 355–602.

Gagné, R. M. (1970). *The Conditions of Learning,* second edition. Holt, Rinehart & Winston.

Gagné, R. M. (1974). *Essentials of Learning for Instruction.* Dryden Press.

Garforth, F. W. (1966). *Dewey's Educational Writings.* Heinemann Educational.

Gattegno, C. (1963). *For the Teaching of Mathematics*, Educational Explorers, Reading.

Glenn, J. A. (ed.) (1977). *Teaching Primary Mathematics: Strategy and Evaluation.* Harper & Row.

Godfrey, C. & Siddons, A. W. (1931). *The Teaching of Elementary Mathematics.* Cambridge University Press. (Godfrey's chapters were written about 1911.)

Goodlad, J. I. (1960). Curriculum: the state of the field. *Rev. Ed. Res.* **30**, 185–98.

Goodlad, J. I. (1966). *School, curriculum and the individual.* Blaisdell.

Goodlad, J. I. (1969). Curriculum: the state of the field. *Rev. Ed. Res.* **39**, 367–75.

Goodlad, J. I. (1976). *The Dynamics of Educational Change.* McGraw–Hill.

Goodlad, J. I. & Richter, M. (1969). *The Development of a Conceptual System for Dealing with Problems of Curriculum and Instruction.* Washington, D.C.

Gordon, P. & Lawton, D. (1978). *Curriculum Change in the Nineteenth and Twentieth Centuries.* Hodder & Stoughton.

Griffiths, H. B. & Howson, A. G. (1974). *Mathematics: Society and Curricula.* Cambridge University Press.

Gruber, H. E. & Vonèche, J. J. (eds.) (1977). *The Essential Piaget.* Basic Books.

Hamilton, D. (1976). *Curriculum Evaluation.* Open Books.

Hamilton, D. *et al.* (1977). *Beyond the Numbers Game: A Reader in Educational Evaluation.* Macmillan.

Harlen, W. (1977). A Stronger Teacher Role in Curriculum Development? *J. Curr. St.* **9**, 21–9.

Havelock, R. G. (1970). *A Guide to Innovation in Education.* University of Michigan.

Hayter, R. J. (1980). *The CMP: a case study.* Schools Council, London.

Heidenheimer, A. J. (1974). The Politics of Educational Reform: Explaining Different Outcomes of School Comprehensivisation Attempts in Sweden and West Germany. *Comp. Ed. Rev.* **18**, 388–410.

Herbert, M. (1974*a*). *Overview, Design and Instrumentation,* Extended Pilot Trials of the CSMP, Evaluation Report Series, No. 1–A–1. CEMREL, St. Louis.

Herbert, M. (1974*b*). *External Review of CSMP Materials,* Evaluation Report Series, No. 1–A–2. CEMREL, St. Louis.

Herbert, M. (1974*c*). *Final Summary Report, Year 1,* Evaluation Report Series, No. 1–A–3. CEMREL, St. Louis.

Hewton, E. (1975). Nuffield Mathematics (5–13): A Profile. *Int. J. Math. Ed. Sci. Technol.* **6**, 407–30.

Hitpass, J. (1967). *Abiturientennachwuchs in der Realschule.* Ratingen bei Düsseldorf.

Hively, W., Maxwell, G., Rabehl, G., Sension, D. & Lundin, S. (1973). *Domain-Referenced Curriculum Evaluation: A Technical Handbook and a Case Study,* CSE Monograph Series in Evaluation, No. 1. Center for the Study of Evaluation, UCLA.

Holz, A., Herbert, M. & Karmos, J. (1974). *Observations of CSMP First Grade Classes,* Extended Pilot Trials of the CSMP, Evaluation Report Series, No. 1–C–2. CEMREL, St. Louis.

Hooper, R. (ed.) (1971). *The Curriculum.* Oliver & Boyd.

House, E. R. (1974). *The Politics of Educational Innovation.* McCutchan.

Howson, A. G. (ed.) (1970). *Developing a New Curriculum.* Heinemann Educational.

Howson, A. G. (1973*a*). Charles Godfrey (1873–1924) and the reform of mathematical education, *Ed. Stud. Math.* **5**, 157–80.

Howson, A. G. (ed.) (1973*b*). *Developments in Mathematical Education.* Cambridge University Press.

Howson, A. G. (1979). A critical analysis of curriculum development in mathematical education. In *New Trends in Mathematics Teaching,* **4**, UNESCO, Paris.

Hudgins, B. (1974). *A catalogue of concepts in the pedagogical domain of teacher education.* Multi-state consortium on performance-based education. Albany, N.Y.

Huhse, K. (1968). *Theorie und Praxis der Curriculumentwicklung,* Max Planck Institut für Bildungsforschung, Berlin.

Jacobsen, E. (1978). Evaluation of Mathematics and Integrated Science Programmes. In *The Evaluation of Integrated Science and Mathematics Curriculum Projects in Africa* (Workshop Report: Seychelles, 1978). UNESCO, Paris.

Jenkins, D. (1976). Curriculum Evaluation. *Curriculum Design and Development,* Units 19–21. Open University Press.

Jivén, L. M. & Öreberg, C. (1968). The IMU Project: Preliminary Plan for Investigating the Effects of a System for Individualized Mathematics Teaching. *Didakometry,* Malmö, School of Education, No. 22.

Kaner, P. (1973). Mathematics for the Majority Project. In *Evaluation in Curriculum Development: Twelve Case Studies.* Schools Council.

Kapadia, R. (1974). A Critical Examination of Piaget–Inhelder's View on Topology, *Educ. Stud. Math.,* **5,** 419–24.

Keitel, C. (1975). *Konzeptionen der Curriculumentwicklung im Bereich des Mathematikunterrichts.* IDM, Bielefeld.

Keitel, C. (ed.) (1980). *Bildung in der Bundesrepublik Deutschland.* Max Planck Institut für Bildungsforschung, Berlin.

Kilpatrick, J. (1979). Methods and results of evaluation with respect to mathematics education. In *New Trends in Mathematics Teaching,* **4,** UNESCO, Paris.

Klein, F. (1939). *Elementary Mathematics from an Advanced Viewpoint (1908).* Dover.

Kline, M. (1961). Math teaching assailed as peril to US scientific progress. *New York University Alumni News,* October 1961.

Kline, M. (1973). *Why Johnny Can't Add: The Failure of the New Math.* St Martin's Press, 1973. (See also Begle, 1974.)

Koerner, J. D. (1968). *Reform in Education: England and the United States.* Delacorte Press.

Krulik, S. & Weise, I. B. (1975). *Teaching Secondary School Mathematics.* W. B. Saunders Co.

Lakatos, I. (1976). *Proofs and Refutations.* Cambridge University Press.

Lalor, J. *et al.* (1839). *The Educator.* Taylor & Walton.

Larsson, I. (1973). *Individualized Mathematics Teaching.* Gleerup, Lund.

Lawton, D. (1973). *Social Change, Educational Theory and Curriculum Planning.* University of London Press.

Lawton, D. (1975). *Class, Culture and Curriculum,* Routledge & Kegan Paul.

Lawton, D. *et al.* (1978). *Theory and Practice of Curriculum Studies.* Routledge & Kegan Paul.

Leithwood, K. A. *et al.* (1976). *Planning Curriculum Change.* Ont. Inst. Stud. Ed.

Lindvall, C. M. & Cox, R. C. (1970). *Evaluation as a Tool in Curriculum Development.* AERA Monograph Series on Curriculum Evaluation, No. 5. Rand McNally.

Lipson, J. I. (1974). IPI Math – An Example of What's Right and Wrong with Individualized Modular Programs. *Learning,* **2,** (8), 60–61.

Long, R. S., Meltzer, N. S. & Hilton, P. J. (1970). Research in mathematics education. *Educ. Stud. Math.* **2,** 446–68.

Lortie, D. (1975). *Schoolteacher: A Sociological Study.* University of Chicago Press.

Lunzer, E. A. (1976). Towards an epistemological theory of mathematics learning. Typescript, University of Nottingham.

Macdonald, J. B. (1967). An example of Disciplined Curriculum Thinking, *Theory into Practice,* **6,** 166–71.

Macdonald, J. B. *et al.* (eds.) (1965). *Strategies for Curriculum Development.* Columbus, Ohio.

McIntosh, A. (1979). When will they ever learn? *Maths. Teaching,* **86,** I–IV.

McLeod, G. K. & Kilpatrick, J. (1969). *Patterns of Mathematics Achievement in Grades 7 and 8: Y-Population.* NLSMA Reports, No. 12. SMSG, Stanford.

Maclure, S. (1968). *Curriculum Innovation in Practice.* HMSO.

Maclure, S. (1971). *Innovation in Education–Sweden.* OECD, Paris.

Maclure, S. (1972). *Styles of Curriculum Development.* OECD, Paris.

Mallinson, V. (1966). *Comparative Education.* 3rd Edn. Heinemann Educational.

Malpas, A. J. (1974). Objective and cognitive demands of the SMP main school course. *Maths. in School,* **3,** (5), 2–5; (6), 20–21.

Maslova, G., Kuznetsova, L. & Leont'eva, M. (1977). Improve the Teaching of Mathematics. *Soviet Ed.* **19** (7), 91–106.

Matthews, G. (1976). Why teach mathematics to most children? *Int. J. Math. Educ. Sci. Technol.* **7,** 253–5.

Mies, Th. *et al.* (1975). *Probleme und Tendenzen der Mathematiklehrerbildung.* IDM, Bielefeld. (Also available in French and English translation.)

Miller, R. L. (1976). Individualized instruction in mathematics: a review of research. *Math. Teacher,* **69,** 345–51.

Morgan, J. (1977). *Affective Consequences for the Learning and Teaching of Mathematics of an Individualised Learning Programme,* Report No. 1. DIME, Stirling University.

Mosher, E. K. (1977). Education and American Federalism: Intergovernmental and National Policy Influences. In *The Politics of Education,* ed. J. D. Scribner. University of Chicago Press.

Neagly, R. L. & Evans, N. D. (1967). *Handbook for Effective Curriculum Development.* Prentice Hall.

Nisbet, J. (1976). Contrasting Structures for Curriculum Development: Scotland and England. *J. Curr. Stud.* **8,** 167–170.

Nyerere, J. (1968). Education for Self-Reliance. In *Ujamaa – Essays on Socialism.* Oxford University Press, Dar-es-Salaam.

Oettinger, A. G. (1969). *Run, Computer, Run.* Harvard University Press.

Orlosky, D. E. & Smith, B. O. (1978). *Curriculum Development: Issues and Insights.* Rand McNally.

Ormell, C. P. (1972). Mathematics-applicable versus pure-and-applied. *Int. J. Math. Ed. Sci. Technol.* **3,** 125–31.

Ormell, C. P. (1974). Bloom's Taxonomy and the Objectives of Education. *Ed. Research,* **17,** 3–18.

Osborne, A. R. (ed.) (1975). Special Issue: Critical Analyses of the NLSMA Reports. *Investigations in Math. Ed.* **8** (3).

Otte, M. (1979). The education and professional life of mathematics teachers. In *New Trends in Mathematics Teaching,* **4,** UNESCO, Paris.

Ovsiew, L. (1973). Research for Better Schools, Inc. (RBS), United States. In *Case Studies of Educational Innovation: I At the Central Level.* OECD.

Owen, J. (1973). *The Management of Curriculum Development.* Cambridge University Press.

Papy, F. (1969). Minicomputer. In *Proceedings 1st ICME,* Reidel, pp. 201–13.

Papy, G. (1963–7). *Mathématique-moderne 1–6,* Didier. (Some volumes are published in translation by Collier-Macmillan.)

Parlett, M. & Hamilton, D. (1977). Evaluation as Illumination: A New Approach to the Study of Innovatory Programmes. In *Beyond the numbers game: a reader in Educational Evaluation,* ed. D. Hamilton. Macmillan.

Piaget, J. (1971). *Science of Education and the Psychology of the Child* (trs. D. Coltman). Longman.

Popham, W. J. (1968). Curriculum materials. *Rev. Ed. Res.* **39**, 319–38.

Popham, W. J. *et al.* (eds.) (1969). *Instructional Objectives*. Rand McNally.

Räde, L. (ed.) (1975). *Statistics at the School Level*. Almqvist & Wiksell Intl.

Revuz, A. (1978). Change in mathematics education since the late 1950s – France. *Ed. Stud. Math.* **9**, 171–81.

Robinsohn, S. B. (1969). *Bildungsreform als Revision des Curriculum*, 2nd Edn. Neuwied am Rhein.

Robinsohn, S. B. & Kuhlmann, J. C. (1967). Two Decades of Non-Reform in West German Education. *Comp. Ed. Rev.* 311–30.

Rosenau, F. (1969). SCIS, AAAS, ESS, IDP . . . Which is Right for Me? *Science Teacher*, **36**, 46–53.

Rosenbloom, P. C. (ed.) (1964). *Modern Viewpoints in the Curriculum*, McGraw Hill.

Ruddock, J. & Kelly, P. (1976). *The Dissemination of Curriculum Development*. NFER.

Sampson, G. (1975). A Personal Assessment. In *The Fife Mathematics Project*, ed. D. H. Crawford. Oxford University Press.

Sarason, S. B. (1971). *The Culture of the School and the Problem of Change*. Allyn & Bacon.

Sauvy, J. & Sauvy, S. (1974). *The child's discovery of space*, Penguin.

Sawyer, W. W. (1978). Oscillations in Systems of Mathematical Education. *Bull. IMA*, **14**, 259–62.

Scandura, J. M. (1971). *Mathematics: Concrete Behavioural Objectives*. Harper & Row.

Schaffarzick, J. (1975). Questions and Requirements for the Comparative Study of Curriculum Development Procedures. In *Strategies for Curriculum Development*, ed. J. Schaffarzick & D. H. Hampson. McCutchan.

Schoen, H. L. (1976). Self-paced mathematics instruction: how effective has it been? *Maths. Teacher*, **69**, 352–7.

Schon, D. A. (1971). *Beyond the Stable State*. Penguin.

Scribner, J. D. (ed.) (1977). *The Politics of Education*. NSSE (76th Yearbook Part 2), University of Chicago Press.

Scriven, M. (1967). The Methodology of Evaluation. In *Perspectives of Curriculum Evaluation*. AERA Monograph Series on Curriculum Evaluation, No. 1, Rand McNally.

Shann, M. H. *et al.* (1975). *Student Effects of an Interdisciplinary Curriculum for Real Problem Solving: The 1974–75 USMES Evaluation*, Final Report. Boston University.

Shulman, L. S. & Keislar, E. R. (eds.) (1966). *Learning by Discovery: A Critical Appraisal*. Rand McNally.

Smith, B. O. (1969). *Teachers for the Real World*. AACTE, Washington.

Smith, D. E. (1970). *Rara Arithmetica*. Chelsea. (Reprint.)

Smith, L. M. & Pohland, P. A. (1974). Education, Technology, and the Rural Highlands. In *Four Evaluation Examples: Anthropological, Economic, Narrative and Portrayal*. AERA Monograph Series on Curriculum Evaluation, No. 7, Rand McNally.

Stake, R. E. (1967). The Countenance of Educational Evaluation. *Teachers College Record*, **68**, 523–40.

Stake, R. E. (1977). Description versus Analysis. In *Beyond the numbers game: a reader in Educational Evaluation*, D. Hamilton *et al*. Macmillan Educational.

Stake, R. E. & Gjerde, C. (1974). An Evaluation of TCITY, the Twin-City Institute for Talented Youth, 1971. In *Four Evaluation Examples: Anthropological, Economic, Narrative and Portrayal*. AERA Monograph Series on Curriculum Evaluation, No. 7, Rand McNally.

Stake, R. E., Easley, J. A. *et al*. (1978). *Case Studies in Science Education* (2 vols). Center for Instructional Research and Curriculum Evaluation, University of Illinois.

Steiner, H. G. (1976). Mathematics Curriculum Development in the USA: A Look at the Past Twenty Years. *Zentralblatt für Did. Math.* **(3)**, 136–40.

Stenhouse, L. (1975). *An Introduction to Curriculum Research and Development*. Heinemann Educational.

Stewart, W. A. C. & McCann, W. P. (1967). *The Educational Innovators, 1750–1880*. Macmillan.

Stone, M. H. (1965). Goals for School Mathematics. *Maths. Teacher*, **58**, 353–60.

Suppes, P. (1965). On the Behavioral Foundation of Mathematical Concepts. In *Mathematical Learning*, Soc. of Res. in Child Development. University of Chicago Press.

Suppes, P. & Morningstar, M. (1972). *Computer-Assisted Instruction at Stanford University 1966–68: Data, Models and Evaluation of the Arithmetic Program*. Academic Press.

Swetz, F. (ed.) (1978). *Socialist Mathematics Education*. Southampton, USA: Burgundy Press.

Sylvester, D. W. (1970). *Educational Documents, 800–1816*. Methuen.

Taba, H. (1962). *Curriculum Development: Theory and Practice*. Harcourt Brace and World.

Tate, T. (1857). *The Philosophy of Education*, (2nd Edn). Longman.

Taylor, L. C. (1972). *Resources for Learning*. Penguin.

Taylor, P. H. & Johnson, M. (eds.) (1974). *Curriculum Development: A Comparative Study*. NFER.

Tebbutt, M. J. (1978). The Growth and Eventual Impact of Curriculum Development Projects in Science and Mathematics. *J. Curriculum Studies*, **10**, 61–73.

Teschner, W-P. (1973). Malmö, Sweden. In *Case Studies of Educational Innovation: II At the Regional Level*. OECD.

Thom, R. (1971). 'Modern' mathematics: an educational and philosophic error. *Amer. Scientist*, **59**, 695–9.

Thom, R. (1973). Modern mathematics, does it exist? In *Developments in Mathematical Education*, ed. A. G. Howson. Cambridge University Press.

Thornbury, R. E. (1973). *Teachers' Centres*. Darton, Longman & Todd.

Thorndike, E. L. (1922). *The Psychology of Arithmetic*. Macmillan.

Thwaites, B. (ed.) (1961). *On Teaching Mathematics*. Pergamon.

Thwaites, B. (1972). *SMP: The First Ten Years*. Cambridge University Press.

Tyler, R. W. (1949). *Basic Principles of Curriculum and Instruction*. University of Chicago Press.

van Engen, H. (1973). The Education of Elementary Teachers. In *Geometry in the Mathematics Curriculum* (36th Yearbook). NCTM.

Vogeli, B. R. (1976). *The Rise and Fall of the 'New Math'*. Teachers College, Columbia University.

Vollrath, H.-J. (1979). Die Bedeutung von Hintergrundtheorien für die Bewertung von Unterrichtssequenzen. *Math. Unterricht*, **5**, 77–89.

Vormeland, O. (1973). The National Board of Education, Sweden. In *Case Studies of Educational Innovation: I At the Central Level*, OECD.

Walker, D. F. (1973). What Curriculum Research? *J. Curr. Stud.* **5**, 58–72.

Walker, D. F. & Schaffarzick, J. (1974). Comparing Curricula. *Rev. Ed. Research*, **44**, 83–111.

Wallis, P. J. (1967). An early Mathematical Manifesto; John Pell's *Idea of Mathematics, Durham Res. Rev.* **18**, 139–48.

Ward, M. (1979). *Mathematics and the 10-year-old*. Evans/Methuen Educational.

Watkins, R. (1973). *In-service Training*. Ward Lock.
Watson, F. (1913). *Vives on Education*. Cambridge University Press.
Weiss, I. (1978). *Report of the 1977 National Survey of Science, Mathematics and Social Studies Education*. Research Triangle Institute, North Carolina.
Whitehead, A. N. (1929). *Aims of Education*. Williams & Norgate.
Whitehead, A. N. (1933). *Adventures of Ideas*. Cambridge University Press.
Wilson, B. J. (1978). Change in Mathematics Education since the late 1950's – West Indies. *Ed. Stud. Math.*, **9**, 355–79.
Wilson, J. W. (1971). Evaluation of Learning in Secondary School Mathematics. In *Handbook on Formative and Summative Evaluation of Student Learning*. B.S. Bloom *et al*. McGraw–Hill.
Wittmann, E. (1976). *Grundfragen des Mathematikunterrichts*. (4th Edn). Vieweg.
Wood, R. (1968). Objectives in the teaching of mathematics. *Ed. Res.* **10**, 83–98.
Wooton, W. (1965). *SMSG: The Making of a Curriculum*. Yale University Press.
Young, M. (1970). Curricula as Socially Organised Knowledge. In *Knowledge and Control*, ed. M. Young. Collier Macmillan.
Ziman, J. (1976). *The Force of Knowledge*. Cambridge University Press.

INDEX